Environmentally Benign Machining

This book provides essential information on environmentally benign/sustainable machining processes including innovations and developments in conventional machining, considering economy, safety, and productivity. Developments in machine tools, recent research on green lubricants and lubrication techniques, process hybridization, and the role of optimization techniques are discussed. Green machining of difficult-to-machine materials and composites is also explained with attempts towards making electric discharge and electrochemical machining technologies.

Features:

- Covers up to date and latest information on environmentally benign machining technologies.
- Includes current approaches regarding the machinability properties of biomaterials, smart materials, and difficult-to-cut materials.
- Reviews theoretical understanding and practical aspects of using different technological approaches to attain sustainability in machining.
- Includes sustainability aspects for both conventional and modern machining.
- Aids industrial users in the optimum selection of machining parameters with regard to sustainability.

This book is aimed at researchers and professionals in manufacturing and mechanical engineering, and sustainable processes.

Advanced Materials Processing and Manufacturing

Series Editor:
Kapil Gupta

The CRC Press Series in Advanced Materials Processing and Manufacturing covers the complete spectrum of materials and manufacturing technology, including fundamental principles, theoretical background, and advancements. Considering the accelerated importance of advances for producing quality products for a wide range of applications, the titles in this series reflect the state-of-the-art in understanding and engineering the materials processing and manufacturing operations. Technological advancements for enhancement of product quality, process productivity, and sustainability, are on special focus including processing for all materials and novel processes. This series aims to foster knowledge enrichment on conventional and modern machining processes. Micro-manufacturing technologies such as micro-machining, micro-forming, and micro-joining, and Hybrid manufacturing, additive manufacturing, near net shape manufacturing, and ultra-precision finishing techniques are also covered.

Advanced Materials Characterization
Basic Principles, Novel Applications, and Future Directions
Ch Sateesh Kumar, M. Muralidhar Singh and Ram Krishna

Thin-Films for Machining Difficult-to-Cut Materials
Challenges, Applications, and Future Prospects
Ch Sateesh Kumar and Filipe Daniel Fernandes

Advanced Materials Processing and Manufacturing
Research, Technology, and Applications
Amogelang Sylvester Bolokang and Maria Ntsoaki Mathabathe

Advanced Joining Technologies
Edited by Manjaiah M, Shivraman Thapliyal and Adepu Kumar

Nanofinishing of Materials for Advanced Industrial Applications
Edited by Faiz Iqbal, Dilshad Ahmad Khan and Zafar Alam

Environmentally Benign Machining
Edited by Şenol Bayraktar and Sunil Pathak

For more information about this series, please visit: www.routledge.com/ Advanced-Materials-Processing-and-Manufacturing/book-series/CRCAMPM

Environmentally Benign Machining

Edited by Şenol Bayraktar
and Sunil Pathak

CRC Press
Taylor & Francis Group
Boca Raton London New York

CRC Press is an imprint of the
Taylor & Francis Group, an **informa** business

First edition published 2025
by CRC Press
2385 NW Executive Center Drive, Suite 320, Boca Raton FL 33431

and by CRC Press
4 Park Square, Milton Park, Abingdon, Oxon, OX14 4RN

CRC Press is an imprint of Taylor & Francis Group, LLC

ISBN: 9781032403038 (hbk)
ISBN: 9781032403052 (pbk)
ISBN: 9781003352402 (ebk)

DOI: 10.1201/9781003352402

Typeset in Times
by codeMantra

Contents

Preface...ix
About the Editors xi
Contributors xiii

Chapter 1 Introduction to Environmentally Benign Machining
and Its Impacts 1

V. Kavimani, P. M. Gopal, V. Sivamaran and K. R. Sumesh
1.1 Introduction 1
1.2 Machining Process Flow in Industrial Sector.......2
1.3 Environment Friendly Manufacturing and Its
Influence Factors 5
    1.3.1 Product Development 6
    1.3.2 Product Processing 6
    1.3.3 Practices for an Eco-Friendly Environment........7
        1.3.3.1 Energy and Resource Consumption in
Machining Operation....................8
        1.3.3.2 Water Pollution and Waste Water in
Machining Processes 9
        1.3.3.3 Airborne Emissions in Product
Machining........................ 10
        1.3.3.4 Solid Waste Materials in Machining
Process 11
    1.3.4 Minimizing Environmental Waste in Machining 11
        1.3.4.1 Recycling and Reuse of Cutting
Fluid.................................... 12
1.4 Summary and Future Development........................ 13
References 14

Chapter 2 Hybrid Advanced Machining for Sustainability in Manufacturing... 17

Şenol Şirin
2.1 Introduction 17
2.2 Details on Hybrid Advanced Machining (HAM) 19
    2.2.1 Laser-Assisted Machining (LAM)...............20
    2.2.2 Vibration-Assisted Machining (VAM).............22
    2.2.3 Heat-Assisted Machining (HAM)...............24
    2.2.4 Electro Chemical Machining (ECM).............25
    2.2.5 Hybrid Water Jet Machining (HWJ) 27
2.3 Summary 28
References 30

Chapter 3   Wire EDM and Laser Beam Machining of NiTi Alloys
for Overall Sustainability 35

M. Adam Khan, G. Ebenezer and J. T. Winowlin Jappes
3.1   Introduction 35
3.2   NiTi Alloy Machining Methods 38
3.3   Wire Electric Discharge Machining (WEDM) 39
3.4   Laser Beam Machining (LBM) 44
3.5   Conclusion 48
References 48

Chapter 4   Eco-Friendly Dielectrics for EDM and WEDM Processes.................54

Mustafa Günay and Ahmet Tolunay Işık
4.1   Introduction 54
4.2   Electrical Discharge Machining (EDM) 55
4.3   EDM Methods 56
       4.3.1   Die-Sinking EDM 57
       4.3.2   Wire EDM 59
4.4   Impacts of Dielectrics on EDM Performance 59
4.5   Eco-Friendly Dielectrics 61
4.6   Recent Studies on EDM and WEDM 66
       4.6.1   Die-Sinking EDM 66
       4.6.2   Wire EDM 68
4.7   Conclusions and Future Research 69
References 70

Chapter 5   Green Machining of Ceramics 78

Ergün Ekici and Şenol Bayraktar
5.1   Introduction 78
5.2   Green Techniques for Conventional Machining
       of Ceramics 79
5.3   Green Techniques for Non-Conventional Machining
       of Ceramics 84
5.4   Conclusions 89
References 90

Chapter 6   Life Cycle Engineering and Analysis of Machining
Processes 96

Suleyman Cinar Cagan
6.1   Introduction 96
6.2   Life Cycle Engineering (LCE) 97
       6.2.1   Definition, Historical Development, and General
                Principles 98
       6.2.2   Stages and Types of LCE 99
       6.2.3   Sustainability and LCE 101

6.3   Machining and LCE 102
       6.3.1   Effect of Machining Parameters on LCE 102
                6.3.1.1   Machining Process Parameters 102
                6.3.1.2   Machining Environments 105
                6.3.1.3   Environmental Effects of Coolant
                           Functions and Strategies 110
6.4   Case Studies 111
       6.4.1   Case 1 111
       6.4.2   Case 2 111
       6.4.3   Case 3 112
       6.4.4   Case 4 113
       6.4.5   Case 5 114
       6.4.6   Case 6 114
       6.4.7   Case 7 114
       6.4.8   Case 8 115
6.5   Summary 115
References 116

Chapter 7   Sustainable Machining of Difficult-to-Machine Materials 121

Emine Şirin
7.1   Introduction 121
7.2   Sustainable or Green Machining Techniques 122
       7.2.1   Dry Machining 123
                7.2.1.1   Dry Machining Case Study 1 123
       7.2.2   MQL Machining 123
                7.2.2.1   MQL Machining Case Study 1 126
       7.2.3   NanoMQL Machining 126
                7.2.3.1   NanoMQL Machining Case Study 1 128
                7.2.3.2   Hybrid NanoMQL Machining 129
                7.2.3.3   Hybrid NanoMQL Machining Case
                           Study 1 129
       7.2.4   Cryogenic Machining 131
                7.2.4.1   Cryogenic Machining Case Study 1 135
       7.2.5   Hybrid Cooling/Lubrication Machining 135
                7.2.5.1   Hybrid Cooling/Lubrication Machining
                           Case Study 1 136
7.3   General Compression of Sustainable Machining 137
7.4   Conclusion 138
References 140

Chapter 8   Environmentally Benign Machining of Composites 143

P. M. Gopal, V. Kavimani and S. Sudhagar
8.1   Introduction 143
8.2   Composites and Their Machining 145
       8.2.1   Composite Materials 145

8.2.2   Machining 146
8.2.3   Challenges in Machining of Composites 147
8.3   Environmentally Benign Machining of Composites............ 148
8.3.1   Dry Machining... 148
8.3.2   Cryogenic Machining... 149
8.3.2.1   Cryogenic Machining of Metal Matrix
Composites.. 149
8.3.2.2   Cryogenic Machining of Polymer
Matrix Composites 150
8.3.3   Machining of Composites with Bio-Based Oils....... 152
8.4   Summary 153
References 154

Chapter 9   Recent Developments in Environmentally Benign Machining........ 157

Mustafa Günay, Ramazan Çakıroğlu and Ahmet Tolunay Işık
9.1   Introduction 157
9.2   Environmental Effects of Machining.................................... 158
9.3   Environmentally Benign Cooling and Lubrication
Systems.. 161
9.4   Recent Research Attempts... 164
9.4.1   Conventional Machining.. 164
9.4.2   Recent Work in Unconventional Machining 170
9.5   Energy Consumption in Machining 172
9.6   Conclusions and Future Work 173
References 174

Index.. 181

Preface

The adaption of sustainability in 4th industrial revolution urges all sectors, including manufacturing to develop and adopt green and environmentally friendly methods and techniques in order to stay competitive in the global economy. The manufacturing sector is one of the most significant contributors in the world economy and always busy promoting research, development, and innovations to meet the accelerated demand for productivity, quality, and sustainability. Intelligent manufacturing processes and techniques have always been helpful to attain that. Therefore, it is important for the engineers, managers, researchers, and professors working in the manufacturing field to understand the development, implementation, and effectiveness of various green and sustainable methods and techniques.

This book facilitates them by providing fundamental understanding, basic knowledge, and advanced research insights into green and environmentally friendly methods and techniques for a wide range of conventional and advanced-type machining processes.

This book consists of *Nine* chapters on Environmentally Benign Machining. Chapter *1* presents a detailed introduction to environmentally benign machining. Chapter *2* sheds lights on the green and hybrid conventional machining technologies for sustainability and their usefulness in current industrial context. Chapter *3* presents a review on machining stratagems of nickel titanium (NiTi) alloy using wire electrical discharge machining and laser beam machining technique, as an example to present their capabilities to achieve sustainability in machining of difficult-to-machine materials. Chapter *4* highlights the eco-friendly dielectrics for electrical discharge machining and wire electric discharge machining Processes, and the effectiveness of using eco-friendly dielectrics has been discussed. Chapter *5* discusses an investigation on green machining of ceramic materials and their effectiveness in industries. Chapter *6* presents insights into life cycle engineering and analysis of various machining processes. This chapter also includes case studies for better understanding of life cycle engineering concept. Chapter *7* presents an interesting study on green machining of difficult-to-machine materials. Chapter *8* highlights the emerging needs for composites and their machining is widely discussed with focus on the environment friendly machining of composite materials. The book ends with Chapter *9* discussing the recent developments in environmentally benign machining, where all the emerging techniques and approaches were briefly highlighted.

We sincerely acknowledge CRC press for this opportunity and their professional support. We are also thankful to all chapter contributors for their availability and valuable contributions.

Şenol BAYRAKTAR and Sunil Pathak

About the Editors

Şenol Bayraktar is an Assoc. Prof. Dr. at Recep Tayyip Erdoğan University, Rize, Turkey. He received his MSc and PhD from Gazi University, Department of Manufacturing Engineering, Ankara, Turkey, in 2011 and 2016, respectively. His research interests involve on the machinability of materials (drilling, turning, milling, etc.) and the efficiency of asynchronous motors. He experienced on the industrial design of neonatal and intensive care devices in the industrial sector and served as manufacturing manager in different sectors. Especially in recent years, he has been working on the machinability of heat treated and non-heat-treated as-cast and additive manufactured aluminum-silicon alloys. He has been working as a manager and researcher in national projects and reviewer in prestigious international journals. He has articles and book chapters published by international publishers.

Sunil Pathak, was awarded his PhD (2016) in Mechanical Engineering from IIT Indore (MP) India and ME and BE in 2012 and 2009 respectively from Rajeev Gandhi Technical University. He is currently an MSCA International Mobility Fellow and working as a Senior Researcher at HiLASE Centre, Czech Republic since Jan 2021. Before joining HiLASE, Sunil has served as a Senior Lecturer at University Malaysia Pahang during Feb 2017–Feb 2020. Presently he is working on laser shock peening of intricately shaped geometries developed by additive manufacturing and improving the cylindrical gear tooth profile by electrochemical machining (ECM). He has authored more than 25 Research articles, 1 monograph, 2 edited books, and more than 15 manuscripts published as book chapters and in various international conferences

Contributors

Adam Khan, M.
Department of Mechanical Engineering
Kalasalingam Academy of Research
  and Education
Krishnankoil, Virudhunagar Dist., India

Bayraktar, Şenol
Department of Mechanical Engineering
Faculty of Engineering and Architecture
Recep Tayyip Erdoğan University
Rize, Türkiye

Cagan, Suleyman Cinar
Department of Mechanical Engineering
Faculty of Engineering
Mersin University
Mersin, Türkiye

Çakıroğlu, Ramazan
Faulty of Technology
Gazi University
Ankara, Türkiye

Ebenezer, G.
Department of Mechanical Engineering
Kalasalingam Academy of Research
  and Education
Krishnankoil, Virudhunagar Dist., India

Ekici, Ergün
Department of Industrial Engineering
Faculty of Engineering
Çanakkale Onsekiz Mart University
Çanakkale, Türkiye

Gopal, P. M.
Department of Mechanical Engineering
Karpagam Academy of Higher
  Education
Coimbatore, Tamil Nadu, India

Günay, Mustafa
Faculty of Engineering
Department of Mechanical Engineering
Karabük University
Karabük, Türkiye

Işık, Ahmet Tolunay
Faculty of Engineering
Department of Mechanical Engineering
Karabük University
Karabük, Türkiye

Kavimani, V.
Department of Mechanical Engineering
Karpagam Academy of Higher Education
Coimbatore, Tamil Nadu, India

Şirin, Emine
Department of Machine and Metal
  Technologies
Duzce University
Düzce, Türkiye

Şirin, Şenol
Department of Mechatronics Engineering
Faculty of Engineering
Duzce University
Duzce, Türkiye

Sivamaran, V.
Department of Mechanical Engineering
Audisanakara College of Engineering
  and Technology (Autonomous)
Gudur, India

Sudhagar, S.
Mechanical Engineering
Kalaignar Karunanithi Institute of
  Technology
Coimbatore

Sumesh, K. R.
Faculty of Mechanical Engineering
Department of Materials Engineering
Czech Technical University in Prague
Karlovo Namesti, Prague, Czech Republic

Winowlin, Jappes J. T.
Department of Mechanical Engineering
Kalasalingam Academy of Research
   and Education
Krishnankoil, Virudhunagar Dist., India

1 Introduction to Environmentally Benign Machining and Its Impacts

V. Kavimani, P. M. Gopal,
V. Sivamaran and K. R. Sumesh

1.1 INTRODUCTION

The movement toward environmentally friendly manufacturing seeks to address the challenge of balancing the need for a thriving global economy with the imperative to protect our planet from further harm. Industrial sectors are driven by the demand for cost-effective production of popular goods while striving to maintain market share. Followed by increased profits for their shareholders, they must find ways to respond to the growing public demand for environmentally safe practices. This is ultimately a question of balancing the needs of the present with those of the future [1, 2]. For manufacturers committed to sustainability, the challenge is developing business strategies that account for economic and environmental considerations. The central challenge of environmentally conscious manufacturing is to reconcile business objectives with environmental imperatives. The key question is how to produce goods that are competitive in the marketplace while avoiding harm to our planet's air, water, and soil. One important consideration is how to motivate companies to adopt environmentally responsible manufacturing practices independently. Additionally, whether nations will take unilateral action to execute environmental standards for producing goods within their borders remains to be seen. While some progress has been made on all of these fronts, there is still far to go before we can protect our ecosystem from negative impact of global manufacturing practices [3, 4].

Environmentally benign machining refers to a set of manufacturing processes that reduce or eliminate the negative environmental impact of traditional machining processes. The societal impact of environmentally benign machining is significant and far-reaching. One of the primary benefits of environmentally benign machining is the reduced environmental footprints. Traditional machining processes generate a significant amount of waste and pollutants that can contaminate soil, air, and water. By reducing the use of harmful chemicals and cutting fluids, environmentally benign machining helps protect the environment. The use of hazardous materials and chemicals in traditional machining processes can lead to health problems for

DOI: 10.1201/9781003352402-1

workers in the manufacturing industry [5, 6]. With environmentally benign machining, hazardous chemicals and materials are minimized, reducing the risk of exposure and related health problems. The focus on environmentally benign machining is a step toward more sustainable manufacturing practices that leads to benefit people, profit, and planet. By reducing the impact of manufacturing procedures, that industry can operate more sustainably, reducing the depletion of natural resources, and minimizing waste. The adoption of environmentally benign machining can lead to economic benefits for manufacturers. Companies that adopt sustainable manufacturing practices are often more attractive to customers and investors who value sustainability. Additionally, manufacturers can save money on materials and energy costs by reducing waste and increasing efficiency. The adoption of environmentally benign machining can have a global impact, as manufacturing is a significant contributor to greenhouse gas emissions. By reducing the environmental impact of manufacturing, we can work toward a more sustainable future for the planet that draws a positive societal impact [7, 8].

While machining is critical to many industrial processes, it can also have a reverse effect on the environment due to significant amount in waste generated and energy consumption. As a result, there is growing interest in environmentally benign machining (EBM), which seeks to decrease the impact machining while maintaining or improving productivity. The goals of EBM are twofold: first, to decree the environmental impact on machining, and second, to improve sustainability of the manufacturing process. EBM seeks to achieve these goals by minimizing energy consumption, reducing the amount of material waste generated, and using environmentally friendly materials [9, 10].

One of the key ways to achieve EBM is through the use of new machining technologies and processes that minimize material waste and energy consumption. For example, high-speed machining approach is used to minimize machining time and energy utilization, although enhance surface finish and dimensional accuracy. Similarly, precision machining techniques, such as laser cutting and electrochemical machining, can be used to reduce material waste and energy consumption by removing material only where it is needed [11, 12]. Another approach to EBM is to use materials that are environmentally friendly. This includes using recyclable, biodegradable materials, or has a reduced environmental impact during production. For example, some manufacturers are using vegetable-based cutting oils that are less environmentally harmful than traditional petroleum-based oils. In addition to using new machining technologies and materials, EBM also involves optimizing the machining process to reduce environmental impact. This includes minimizing the amount of coolant and lubricant used during machining, reducing the amount of energy consumed during machining, and optimizing the machining process to reduce waste [13, 14].

1.2   MACHINING PROCESS FLOW IN INDUSTRIAL SECTOR

In the modern era, technological advancements have allowed us to manufacture goods at peak levels, successfully overcoming shortages in numerous fields by mass production. However, this fabrication of different materials has resulted in mass pollution and the release of large amounts of solid waste into the environment. This waste solid must be reutilized in a secondary cycle of manufacturing in various categories. Industrial sectors such as fertilizer, mining pesticides, and cement has major contributed the most to pollution over the past few decades. The demand for cement and concrete has increased dramatically in recent decades, and these industries alone account for 5% of greenhouse gases. As a result, it is necessary to control pollution in these industries by changes in the fabrication process of concrete and cement. When production articles made by ferrous and non-ferrous metals. Scrap is produced that is difficult to recycle by conservative methods [15]. Hence, there is a need for economic and environmentally friendly methods for reusing the scrap without melting it, due to its higher cost. Due to innovative development, water requirements have increased abruptly in industrial and urban areas in recent decades. However, 15%–20% of water utilized for human uses, while the rest, 80%–85%, is utilized for industrial growth, negatively affecting natural water resources. Although there have been successful attempts to purify water bodies, the capital cost of the production system has increased significantly [16, 17]. Every production process must aim to minimize waste and minimize its reverse impact over environment. To attain this, source materials must pass through multiple production processes so each step generates a beneficial product with minimal environmental damage.

EBM encompasses a range of technologies, practices, and methods that support sustainable production within the industrial ecology framework. EBM aims to achieve zero waste by developing and implementing benign materials processing strategies that facilitate reuse and use environment. Additionally, EBM addresses the challenges of total environmental management, including remanufacturing, reuse, and recycling. A recent study by the World Technologies Evaluation Center found Europe, Japan, and the United States had different approaches to addressing environmental impacts; they shared a common goal of reducing such impacts. Based on 52 site visits, the World Technologies Evaluation Center panel concluded that a complete life cycle analysis is essential to evaluate decisions that reduced impact on environmental. Figure 1.1 shows the schematic cycle in manufacturing industries. It refers to extracting, using, and recycling materials within industrial processes.

This cycle includes the extraction of raw materials, the processing of these materials into products, and the eventual disposal or recycling of these products. The industrial material cycle is typically broken down into three stages: extraction, production, and disposal. In the extraction stage, raw materials such as metals, minerals, and fossil fuels are extracted from the earth. These materials are then transported to production facilities and processed into finished products such as cars, electronics, and buildings. In the production stage, these finished products are used and eventually reach the end of their useful life. At this point, they are either disposed of or recycled. Disposal can take the form of landfilling, incineration, or other waste management methods. Recycling involves collecting and processing used materials to be used again in new products [18, 19]. Efficient and sustainable industrial material cycles rely on reducing, reusing, and recycling. This means reducing the amount of materials used in production, reusing materials whenever possible, and recycling materials at the end of their useful life. By implementing these principles, industries can reduce

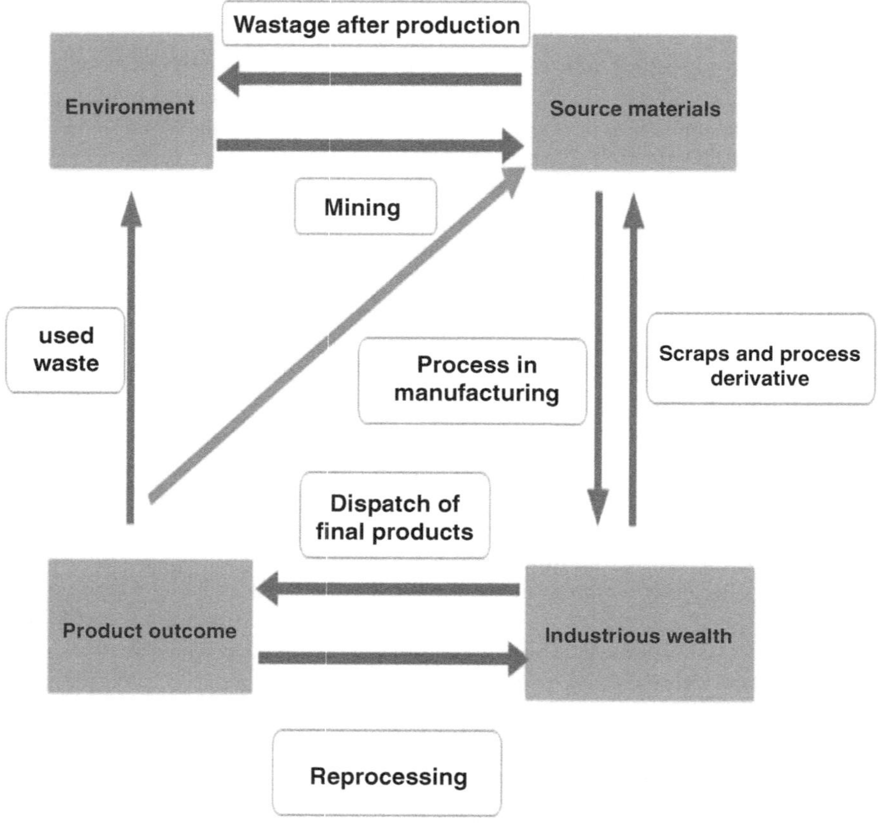

FIGURE 1.1   Schematic view on material cycle in manufacturing industry

their environmental impact and create a more circular economy. Figure 1.2 illustrates material cycle in industrial known as product development life cycle. This framework is beneficial for analyzing the product development stage, which involves acquiring materials, processing them, maintaining the product, and considering options for reuse, recycling, and packaging. Although the framework simplifies the complex connectivity of these aspects, it highlights the issues that need to be addressed when evaluating alternatives to current manufacturing practices. Moreover, it helps to identify opportunities for reducing the environmental impact of current practices by targeting the most accessible areas for improvement [20, 21]. Various industries have developed similar frameworks and complemented them with matrix methods or databases that organize product and process assessments. However, this representation does not consider factors such as energy and water usage or the generation of other types of waste like solvents and cutting fluids.

Net-shape manufacturing involves a substantial amount of the conversion from the coil to product, but other cycle elements, such as extraction of ore, production in stock metal from ore and waste, and product is crucial for a thorough life-cycle study.

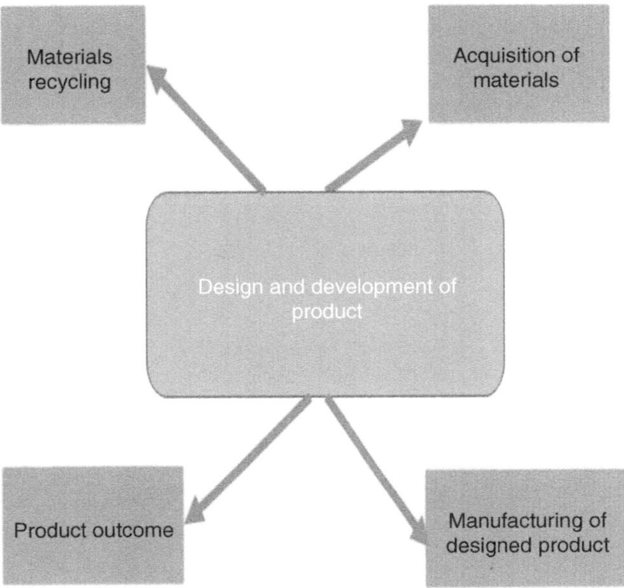

FIGURE 1.2 Schematic view on product life cycle

However, obtaining complete information on the equipment and material transformation processes might be challenging.

Environmentally benign manufacturing is essentially the same as the general manufacturing process. The only difference is that it includes measures to minimize the amount of scrap and emissions released into the surrounding environment. To achieve this, waste materials are processed through a secondary manufacturing cycle and pollution control methods are employed to regulate harmful emissions [20, 22]. This method is referred to as environmentally benign manufacturing because it provides extra provisions for eco-friendly and cost-effective manufacturing. The current market demands industries like pesticides and mining, which successfully fulfill these demands but contribute significantly to particulate emissions. Additionally, India's steel and aluminum industries are experiencing peak production and require secondary scrap manufacturing. EBM includes provisions for reusing and recycling waste consumption, as depicted in Figure 1.3.

1.3 ENVIRONMENT FRIENDLY MANUFACTURING AND ITS INFLUENCE FACTORS

In the past, production process comparisons for various matrixes of products were factored into manufacturing strategy. Nowadays, manufacturing strategies frequently mix organizational and philosophical elements by considering both processes and products. The association among environment and production processes and the accelerated demands due to increased population are being concerned these days.

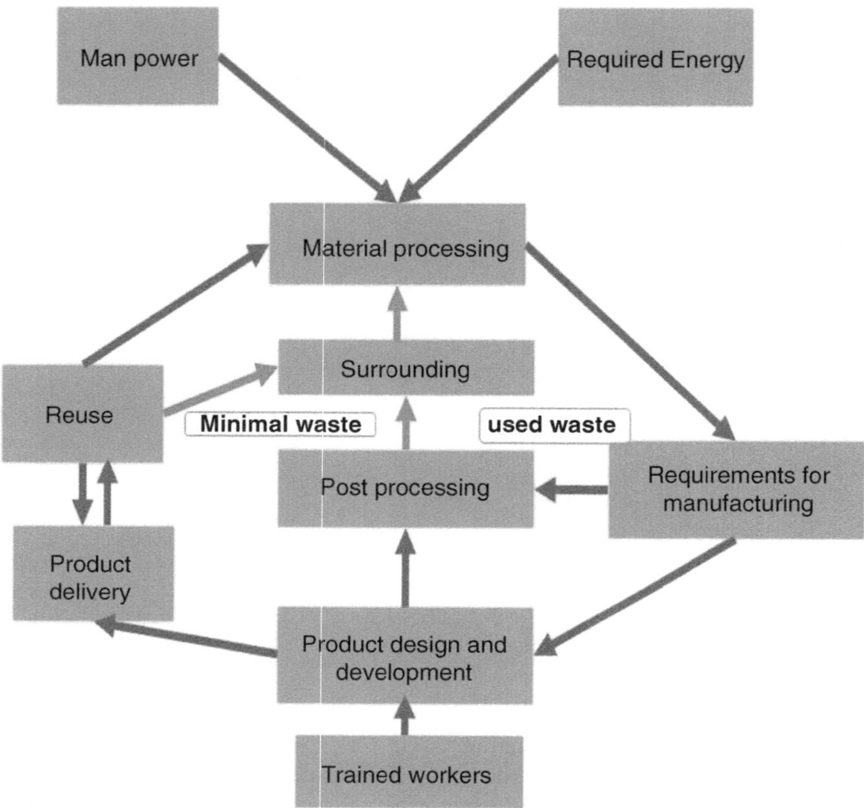

FIGURE 1.3   Representation of environmentally benign manufacturing

The following three factors have an effect on the technology category related EBM [23].

1.3.1 PRODUCT DEVELOPMENT

When designing environmentally friendly products, manufacturers usually consider the effects on the environment throughout the product's life. It commonly mentions the application of life cycle analysis and design for the environment methods as a consequence. Environmentally responsible product design may help introduce and maintain new products. Product flexibility, for instance, enables environmental improvements like source material substitution while preserving competitiveness.

1.3.2 PRODUCT PROCESSING

Environmental advantages in manufacturing processes are linked to reduction, reuse, recycling, and remanufacturing. The recycling of wastage is required by zero-emission manufacturing, which views the production system as industrial eco-system. Hence, expertise in recycling and reduction in pollution is required in manufacturing with no emissions. Manufacturing equipment that can adapt to shifting material flows can assist to increase sustainability while maintaining competitiveness. The ability to use flexible materials is also necessary for flexible production. For instance, more effective and recyclable materials may be used to make ecologically friendly packaging [24, 25].

1.3.3 PRACTICES FOR AN ECO-FRIENDLY ENVIRONMENT

The ISO 14000 certification has a substantial environmental impact on an organization's manufacturing practices, even while it can only support organizational practices and cannot guarantee environmental benefits on its own. Other tactics that may be strategically used to improve manufacturing include benchmarking. These three components complement one another, are linked, and have certain areas of overlap. The majority of technology advances, result in collaboration among numerous firms, usual support of government [26].

This is true even if technological improvements can emerge from within a company. Manufacturing approaches viz., drilling, turning, milling (Figure 1.4), thread cutting, boring and grinding are the better illustrations for machining types. Each, procedure entails metal removing from raw source or work samples that is various form of cubical block, cylindrical bar or pellet stock. The goal of metal-cutting economics is to maximize one of the three basic factors: product output, tool life and operation cost for materials operations. These manufacturing approaches decrease the environmental effect that is list on the economic goals. Herein lubricants and coolants are the primary needs in machining process, fulfilling these industrial goals' needs in machining [27]. These lubricants and coolants have a vital role in EBM as follows:

- Reducing frictional heat
- Decrement in heat-related component deformation
- minimizing friction between chips, the tool, and the workpiece
- Cleaning up chips,
- Minimizing potential corrosion on the machine and the work item
- Avoiding the product or part's built-up edges

The two main environmental effects of machining processes are the collection of metal chips and the discharge of cutting fluids into the environment. Recycling chips by using them as a charge in the metal-casting process is the best answer to the issue of chip buildup. However, recycling could require moving the chips to a far-off location, which would have a significant impact. Recycling cutting fluids back into the machining process is the best approach to handle them. To do this, the chips from the machining effluent must be removed, and the cutting fluid must be reconstituted to a state as close to its initial as is practical [28].

Each of these stages comes at a financial expense that must be weighed against the cost of the environmental effect of just throwing away the chips and wasted cutting

FIGURE 1.4 Vertical machining center

fluid. In some procedures, cutting fluids may also be utilized to provide cooling and lubrication. Hydraulic oil and lubricants are required to keep the machine tools functioning properly. Chips and swarf are generated during the machining process while material is removed. Mists, volatile organic, and dusts compounds may be produced, based on adopted method. The environmental effects of these material and energy fluxes will be examined in the sections that are as follow.

1.3.3.1 Energy and Resource Consumption in Machining Operation

Rising energy costs and worries about global warming have sparked a growing interest in energy efficiency across a range of manufacturing industries. The typical power rating of a contemporary machining center is between a few kilowatts and many tens of kilowatts. A machining center's other energy-consuming parts, viz., coolant system, feed drive, auxiliary equipment, and other control system. There are three categories of power usage: basic, idle, and cutting. In numerous machining operations, cutting fluids are utilized to increase tool life, enhance surface quality, and remove chips. Coolants are often straight oils made of lubricant additives such mineral oil, biocides, etc. In addition to cutting fluids, the machines also use lubrication oils and hydraulic fluids, but in considerably minimal quantities. Cutting inserts and the tools are the major material input in material removal process. In addition to these main material inputs, cutting conditions can also be decisive in determining tool life in cutting tools.

Conversely, manufacture of tools may be energy intensive and have a major negative impact on the environment. Cements, steel, carbides and diamond are characteristic tool insert materials. Tantalum tungsten, boron, and titanium are used to develop cutting tools and inserts; similarly nickel and cobalt are commonly used as binders. The availability of several metals may be limited in near future owing to their low abundance in nature. There are some ways in which minimizing energy and resource consumption can benefit the environment. One of these is optimizing environmental benign machining methods. This can provide assistance to lessen climate change and reduce the environmental impact of machining. Minimizing the amount of resources consumed during machining, environmental benign machining can help conserve natural resources [29, 30]. This can reduce the environmental impact of resource extraction and promote sustainability. Minimizing waste generation during machining, environmental benign machining can reduce the environmental impact of waste disposal. This can help to prevent pollution and reduce the amount of material that ends up in landfills. Reducing the consumption of coolant fluids during machining, environmental benign machining can help to improve water quality. This can reduce the environmental impact of machining on aquatic ecosystems and promote sustainable water use. Overall, minimizing energy and resource consumption through benign environmental machining can significantly impact the environment. By promoting sustainability and reducing ecological effects in machining, environmental benign machining help create more sustainable and resilient future.

1.3.3.2   Water Pollution and Waste Water in Machining Processes

Wastewater and water pollution are significant environmental concerns associated with machining processes. Environmental benign machining seeks to minimize the environmental impact of these processes by reducing wastewater and water pollution. Here are some ways that environmental benign machining can help mitigate the environmental impacts of wastewater and water pollution. One of the main sources of wastewater in machining processes is coolant fluids used to reduce friction and heat during cutting and shaping. Environmental benign machining seeks to minimize the use of these fluids and promote their recycling. This can reduce the amount of wastewater generated during machining and promote sustainable water use. Machining processes can generate wastewater that contains pollutants that can harm the environment. Environmental benign machining can mitigate the impact of these pollutants by implementing water treatment processes that remove harmful substances before the water is discharged into the environment. Some coolant fluids used in machining can contain toxic substances that can harm the environment. Environmental benign machining seeks to use non-toxic coolant fluids that are safer for the environment and reduce the risk of water pollution. Environmental benign machining involves monitoring the quality of wastewater generated during machining processes. By monitoring water quality, manufacturers can identify and mitigate the environmental impacts of their operations. Overall, environmental benign machining can help to mitigate the environmental impacts of wastewater and water pollution associated with machining processes. By promoting sustainable water use and reducing the release of pollutants into the environment, environmental benign machining can help to create a more sustainable and resilient future.

While in operation, coolant oils are pumped back and forth between storage, upkeep, and the machining process. Due to evaporation certain liquids and substances are lost. In order to sustain function, specific compounds are often provided on a regular basis. In addition, fluid contaminants such as lubricants and dust from surroundings will develop. Coolant can support growth of microorganisms because they contain an appropriate ratio of water and organic ingredients. Cutting fluids only have a very short service life, even with maintenance procedures, thus the used cutting fluids must be appropriately handled before disposal. The used cutting fluids contain a variety of hazardous materials. Heavy metals from the workpiece and the cutting as well as coolant components among these. Additionally, the fluids' oil, fat, and grease content causes a large reduction in oxygen and increased nutritional loading [31]. There is currently no ideal technique for treatment of wastewater for variable and complex collection of pollutants, and typical wastewater treatment solutions frequently fall short of regulatory standards. Several treatment methods are chosen depending on how and where the institution operates. Waste cutting fluids are considered hazardous wastes in the European Union (EU) and must be managed as such.

1.3.3.3   Airborne Emissions in Product Machining

Airborne emissions are another significant environmental concern associated with machining processes. Environmental benign machining seeks to minimize the environmental impact of these emissions by reducing the release of pollutants into the air. Here are some ways that environmental benign machining can help mitigate the environmental impacts of airborne emissions. EBM pursues to reduce the energy consumption of machining processes. This can reduce the amount of energy generated from non-renewable sources, which can contribute to the release of pollutants into the air. EBM follows to use equipment that is designed to be more energy efficient. This can reduce the energy required to operate machining processes, reducing the release of airborne emissions. EBM can promote the use of low-emission fuels for equipment that generates emissions, such as generators or forklifts. This can reduce the amount of pollutants released into the air during machining processes. It involves the use of equipment that captures emissions generated during machining processes. These emissions can be treated to remove harmful substances before they are released into the air, reducing the environmental impact of machining [32]. General, EBM can help to mitigate the environmental impacts of airborne emissions associated with machining processes. By reducing energy consumption, promoting efficient equipment and low-emission fuels, and capturing and treating emissions, environmental benign machining can help create a more sustainable and resilient future. The volatile organic compounds and particles are released into air during machining. When machining with cutting fluids, the fluid spray impacts the surfaces of workpiece before breaking off into minute droplets that are transported away by air flows. Additionally, liquids will momentarily cling to revolving surfaces until centrifugal force causes them to disperse as droplets. As large heat is produced droplets can also develop from the vapor of the constituents in cutting fluids. The size of the droplets produced by vapor condensation is often less than that of the droplets produced by centrifugation.

Typically, mist produced through machining differs chemically from the coolant. It could contain some of the compounds found in cutting fluids and pollutants like bacteria and metals. Mist concentrations in the range of 1 to 10 mg/m^3 are typical during milling. The size of the droplets has a considerable impact on human health and the environment in addition to the chemical makeup. Small droplets can float in air for lengthy time and are challenging to collect, whereas large droplets settle down rapidly and are readily removed. They can enter the lungs' alveoli and permeate there, causing irritation, allergic responses, and carcinogenic consequences when breathed. Coolant mists might arrive in body through skin in addition to inhalation, which may result in allergic responses and rashes. Dust and particle emissions are frequent during dry machining, and the dust produced depends on the cutting depth, feed rate, and speed. Airborne emissions are often found to be substantially lower during dry machining than they are during wet machining [33].

1.3.3.4 Solid Waste Materials in Machining Process

It is the form of chips and swarf that are inevitable by-product of machining as it is a material removal operation. The residue produced during grinding processes also includes particles left behind from the grinding wheels. Significant breakthroughs in zero-waste machining have been made possible by tighter landfilling regulations and rising material prices.

The swarf and chips are frequently together and repurposed for the recovery of material. Used cutting tools/inserts are additional solid wastes produced during machining operations such as turning operation (Figure 1.5). These days, returning them to suppliers and manufacturers for remanufacturing and repurposing is standard procedure. Solid waste is one of the environmental impacts associated with machining processes. Environmental benign machining seeks to minimize the generation and disposal of solid waste by adopting sustainable waste management practices. Here are some of the ways that environmentally benign machining can help to mitigate the environmental impacts of solid waste. EBM minimize the material waste by using efficient cutting tools and processes. This can minimize the rate of material needs to be discarded, thus decrease the quantity of solid waste generated during machining processes. EBM promotes recycling and reusing materials, such as scrap metal and coolant, generated during machining processes. This can reduce the amount of solid waste that needs to be disposed of and conserve natural resources. EBM involve implementing sustainable waste management practices, such as sorting and segregating waste, and disposing of waste in an environmentally responsible manner. This can reduce the environmental impact of solid waste generated during machining processes. Biodegradable cutting fluids that are less harmful to the environment. This can reduce the amount of hazardous waste generated during machining processes and help to create a more sustainable and resilient future. The generation and disposal of solid waste by implementing sustainable management practices and using biodegradable cutting fluids. By doing so, environmental benign machining can help to mitigate the environmental impacts of solid waste associated with machining processes [34].

1.3.4 Minimizing Environmental Waste in Machining

Coolant and fluid are used as fluids for cutting in manufacturing operations, which is the main source for environmental issues. Because of this, majority of R&D efforts

FIGURE 1.5   CNC turning center

on machining environmentally friendly process that concentrated in doing away with drastically reducing in usage of traditional fluids with more environment friendly options, there are primarily two categories that may be used for EBM.

- **Coolant-based approach**: This strategy entails methods that lessen the environmental effect of cutting fluids by extending their shelf lives or substituting biodegradable fluids for them. This method does not call for any modifications to the fundamental machine or cutting tool. To recycle and reuse the cutting fluids, several peripheral rearrangements are needed. This section primarily focuses on this fluid-based methodology.
- **Planning approach:** This strategy strives to completely forego or drastically reduce the usage of cutting fluids. The basic functions of coolant fluids are accomplished by altering coolant and tools, namely flushing of chips and lubricating.

1.3.4.1   Recycling and Reuse of Cutting Fluid

Several pollutants will enter cutting fluids during usage, as was previously mentioned. These impurities will reduce machining efficiency and ultimately necessitate the discharge of cutting fluids. Cutting fluids are restored to like-new state when these pollutants are removed, extending their useful life. This will lessen the negative environmental effects of using cutting fluids and the health hazards to personnel exposed to them. The machining procedures and the workpiece material's nature determine the allowable amount of particles. Removing tiny particles is a little trickier, and separation or filtration is frequently required. Due to their low maintenance requirements, disposable media, including bags, cartridges, and wrapped items, are increasingly widely used [35]. However, the method may also remove particles with a smaller diameter because as the filtering process continues, a cake layer forms on the surface. Smaller particles can be captured by this cake layer, albeit at expense of lower infusion flow rate. When the filtering speed drops, the medium must be replenished. Similar steps are taken when using permanent media like woven fabric belts. Indexing is done to eliminate cake from medium by smearing an air when the filtration rate reaches a specific point. This makes it possible to reuse the supporting media for subsequent cycles. Perpetual media often has high pore diameters. The filtered fluid could be substantially less clean before a cake layer forms. Surfactants that are already present in the cutting fluids may be able to loosely emulsify some of the free oils that enter them. A modest amount of free oils, as opposed to particles, might be advantageous since they increase lubricity. Skimmers, coalescers and centrifuges are often used oil removal tools. Additionally, oil sorbent materials are frequently employed. The oil is extracted and composed for reuse. It must be mentioned the microorganisms cannot be eliminated from the fluids by the separation and filtering systems covered above. Recently, the removal of contaminants via membrane microfiltration has received attention. Microfiltration has the benefit of being able to remove particles and free oils in addition to bacteria. Membrane microfiltration is frequently carried out in the crossflow mode, which is filtration parallel to fluid flow. The cake accumulation is regulated by force from the bulk flow in this operating mode, which reduces the clog of membrane pores. The membranes may be cleaned and reused again and are often constructed of ceramic and polymeric materials. By avoiding use of problematic biological control, it improves the environment and workplace safety. Cutlery with a bio-based base Oil and other ingredients mostly sourced from petroleum are traditionally used in the formulation of coolant [36, 37]. Therefore, high human/ecological toxicity and limited biodegradability, many of these substances pose health concerns toward employees that disposal more challenging. Groundwater contamination might all arise from improper cutting fluid disposal. Cutting fluids containing chemicals obtained from biological resources have been developed during the past 20 years in an effort to solve these problems. Canola oil, palm and jatropha oil are a few vegetable oils used to create coolant fluids. Vegetable oil-based cutting fluid formulation probably calls for different surfactant packages since molecular structure differs from other oil's [38].

1.4 SUMMARY AND FUTURE DEVELOPMENT

Excessive cutting fluid usage, waste generation, energy and resource consumption, health and safety related issues, and high cost are the major challenges in machining sector. Most of them directly lead to environmental degradation. Significant attempts have been undertaken to address them during the past several decades. An overview of these initiatives is provided in this chapter. These initiatives might be divided to two groups: fluid and system-based approaches. The useable life of cutting fluids has been extended, and formulations for cutting fluids employing bio-based chemicals have been developed for fluid-based techniques. The use of cutting fluids has been minimized or discontinued for system-based techniques. It should be emphasized that many of these methods will raise manufacturing costs and lack viability in a number of applications. A life cycle evaluation must be done to establish the real environmental advantages of new machining technology because they most likely have their own environmental costs. To deliver timely feedback, life cycle evaluation should ideally be incorporated into process development. Therefore, there is a pressing need to create a thorough life cycle inventory database covering various machining techniques. Environmental sustainability has emerged as one of our society's major challenges as both the global population and the standard of living rise. Environmental effects of manufacturing are important, particularly in poor nations. Therefore, reaching significant sustainable development targets depends on greening industrial processes, including machining operations. The development of green technology in machine shops has challenges that must be solved. The research community should keep working on innovative machining technologies with enhanced economic, societal benefits, and environmental performance at the same time. One of the most significant production techniques used nowadays in business is machining. But a number of environmental problems linked to machining operations have been found. These problems include the use of energy and resources, airborne pollutants to occupational health. The main source of the majority of the problems is the pervasive usage of cutting fluids in machining processes. During the past decades of 20 years, R&D into EBM has mostly concentrated on minimizing the negative effects of employing cutting fluids on the environment. A thorough analysis of these initiatives was given in this chapter. Cutting fluids can be made to last longer by having particles, free oils, and other impurities removed using separation and filtering processes. New bio-based formulations that improve biodegradability and lower toxicity without compromising machining performance were developed to substitute petroleum fluids. Cutting fluid usage can be decreased or even completely eliminated during machining thanks to the formation of innovative materials on tool and coatings. Effective cooling is accomplished in dry and minimal lubrication machining by providing tiny amounts of water vapor, liquid nitrogen, etc. Along with these methods, the machining process may be adjusted to have less of an impact on the environment, particularly in terms of energy use and carbon emissions. Due to economic and knowledge constraints, many novel approaches have only had limited industrial adoption. This problem has to be solved by community of researchers and industry deals with machine tool working together.

REFERENCES

[1] B. Sarkar, S. Bhuniya, A sustainable flexible manufacturing–remanufacturing model with improved service and green investment under variable demand, *Expert Syst. Appl.* 202 (2022) 117154.

[2] D.Y. Pimenov, M. Mia, M.K. Gupta, Á.R. Machado, G. Pintaude, D.R. Unune, N. Khanna, A.M. Khan, Í. Tomaz, S. Wojciechowski, Resource saving by optimization and machining environments for sustainable manufacturing: A review and future prospects, *Renew. Sustain. Energy Rev.* 166 (2022) 112660.

[3] Y. Chen, B. Lin, Towards the environmentally friendly manufacturing industry–the role of infrastructure, *J. Clean. Prod.* 326 (2021) 129387.

[4] L.R. Silva, E.C.S. Corrêa, J.R. Brandao, R.F. de Avila, Environmentally friendly manufacturing: Behavior analysis of minimum quantity of lubricant-MQL in grinding process, *J. Clean. Prod.* 256 (2020) 103287.

[5] T. Kurita, C. Endo, Y. Matsui, H. Masuda, K. Terasawa, F. Tanaka, H. Ikeda, K. Oguchi, K. Kobayashi, Mechanical/electrochemical complex machining method for efficient, accurate, and environmentally benign process, *Int. J. Mach. Tools Manuf.* 48 (2008) 1599–1604.

[6] M.K. Gupta, Q. Song, Z. Liu, C.I. Pruncu, M. Mia, G. Singh, J.A. Lozano, D. Carou, A.M. Khan, M. Jamil, Machining characteristics based life cycle assessment in eco-benign turning of pure titanium alloy, *J. Clean. Prod.* 251 (2020) 119598.

[7] X. Yang, H. Cao, B. Li, S. Jafar, L. Zhu, A thermal energy balance optimization model of cutting space enabling environmentally benign dry hobbing, *J. Clean. Prod.* 172 (2018) 2323–2335.

[8] J.S. Nam, D.H. Kim, H. Chung, S.W. Lee, Optimization of environmentally benign micro-drilling process with nanofluid minimum quantity lubrication using response surface methodology and genetic algorithm, *J. Clean. Prod.* 102 (2015) 428–436.

[9] X.-C. Tan, Y.-Y. Wang, B.-H. Gu, Z.-K. Mu, C. Yang, Improved methods for production manufacturing processes in environmentally benign manufacturing, *Energies.* 4 (2011) 1391–1409.

[10] T. Gutowski, C. Murphy, D. Allen, D. Bauer, B. Bras, T. Piwonka, P. Sheng, J. Sutherland, D. Thurston, E. Wolff, Environmentally benign manufacturing: Observations from Japan, Europe and the United States, *J. Clean. Prod.* 13 (2005) 1–17.

[11] M.H. Cetin, B. Ozcelik, E. Kuram, E. Demirbas, Evaluation of vegetable based cutting fluids with extreme pressure and cutting parameters in turning of AISI 304L by Taguchi method, *J. Clean. Prod.* 19 (2011) 2049–2056.

[12] B. Ozcelik, E. Kuram, E. Demirbas, E. Şik, Optimization of surface roughness in drilling using vegetable-based cutting oils developed from sunflower oil, *Ind. Lubr. Tribol.* 63 (2011) 271–276.

[13] M. Guiton, D. Suárez-Montes, R. Sánchez, P. Baustert, C. Soukoulis, B.S. Okan, T. Serchi, S. Cambier, E. Benetto, Comparative life cycle assessment of a microalgae-based oil metal working fluid with its petroleum-based and vegetable-based counterparts, *J. Clean. Prod.* 338 (2022) 130506.

[14] J.E. Manikanta, B.N. Raju, C. Prasad, B.S.S.P. Sankar, Machining performance on SS304 using nontoxic, biodegradable vegetable-based cutting fluids, *Chem. Data Collect.* 42 (2022) 100961.

[15] A. Singh, A. Anand, J. Singh, M. Singh, K. Kumar, P. Bhandari, V.K. Srivastava, A. Singh, M.K. Sinha, Suitability of turning and grinding steel chips to synthesize metal matrix composite via powder metallurgy route, *Mater. Today Proc.* 69 (2022) 215–219.

[16] A. Singh, H.S. Dhami, M.K. Sinha, R. Kumar, Evaluation and comparison of mineralogical, micromeritics and rheological properties of waste machining chips, coal fly ash particulates with metal and ceramic powders, *Powder Technol.* 408 (2022) 117696.

[17] V. Muralidharan, S. Palanivel, M. Balaraman, Turning problem into possibility: A comprehensive review on leather solid waste intra-valorization attempts for leather processing, *J. Clean. Prod.* 367 (2022) 133021.

[18] M. Soori, M. Asmael, A review of the recent development in machining parameter optimization, *Jordan J. Mech. Ind. Eng.* 16 (2022) 205–223.

[19] J. Zhao, S. Li, Life cycle cost assessment and multi-criteria decision analysis of environment-friendly building insulation materials-A review, *Energy Build.* 254 (2022) 111582.

[20] A.I. Gomez-Merino, J.J. Jiménez-Galea, F.J. Rubio-Hernandez, I.M. Santos-Ráez, Experimental assessment of thermal and rheological properties of coconut oil-silica as green additives in drilling performance based on minimum quantity of cutting fluids, *J. Clean. Prod.* 368 (2022) 133104.

[21] D.N.S. Naik, V. Sharma, Thermophysical investigations of mango seed oil as a novel cutting fluid: A sustainable approach toward waste to value addition, *J. Manuf. Sci. Eng.* 144 (2022) 91004.

[22] N. Khanna, J. Wadhwa, A. Pitroda, P. Shah, J. Schoop, M. Sarıkaya, Life cycle assessment of environmentally friendly initiatives for sustainable machining: A short review of current knowledge and a case study, *Sustain. Mater. Technol.* 32 (2022) e00413.

[23] P.G. Ponnusamy, S. Mani, Life cycle assessment of manufacturing cellulose nanofibril-reinforced chitosan composite films for packaging applications, *Int. J. Life Cycle Assess.* 27 (2022) 380–394.

[24] V. Kannan, Experimental study of an eco-friendly turning process of nimonic 75 Combining minimum quantity lubrication and hexagonal boron nitride-enhanced neem and jatropha oil nanofluids, *J. Inst. Eng. Ser. C.* 103 (2022) 785–812.

[25] R. Bag, A. Panda, A.K. Sahoo, A concise review on environmental sustainable machining conditions of hard part materials, *Mater. Today Proc.* 62 (2022) 3724–3728.

[26] A. Barragán-Ocaña, P. Silva-Borjas, K.A. Luna-López, M.D.P. Longar-Blanco, Technological development and mitigation of environmental impact through ISO 14000: A review, *Int. J. Product. Qual. Manag.* 36 (2022) 27–45.

[27] G. Kshitij, N. Khanna, Ç.V. Yıldırım, S. Dağlı, M. Sarıkaya, Resource conservation and sustainable development in the metal cutting industry within the framework of the green economy concept: An overview and case study, *Sustain. Mater. Technol.* 34 (2022) e00507.

[28] R. Wang, M. Wang, S. Gao, Z. Wang, T. Xin, M. Liu, Y. Bao, Compound purification of nickel base superalloy cutting waste through special cleaning agent attached to ultrasonic stirring, *J. Clean. Prod.* 378 (2022) 134548.

[29] K.Z. Yang, A. Pramanik, A.K. Basak, Y. Dong, C. Prakash, S. Shankar, S. Dixit, K. Kumar, N.I. Vatin, Application of coolants during tool-based machining–A review, *Ain Shams Eng. J.* 14 (2022) 101830.

[30] R.A. Kazeem, D.A. Fadare, O.M. Ikumapayi, A.A. Adediran, S.J. Aliyu, S.A. Akinlabi, T.-C. Jen, E.T. Akinlabi, Advances in the application of vegetable-oil-based cutting fluids to sustainable machining operations—A review, *Lubricants.* 10 (2022) 69.

[31] J. Józwik, M. Zawada-Michałowska, G. Budzik, S. Legutko, M. Kupczyk, Microbiological analysis of coolant used in machining, *Adv. Sci. Technol. Res. J.* 17 (2023) 206–214.

[32] C. Ravasio, G. Pellegrini, M. Quarto, Development of CO_2 efficiency index for evaluating sustainability of microelectrical discharge drilling process, *Proc. Inst. Mech. Eng. Part B J. Eng. Manuf.* 237 (2023) 758–769.

[33] M.Q. Saleem, A. Mehmood, Eco-friendly precision turning of superalloy Inconel 718 using MQL based vegetable oils: Tool wear and surface integrity evaluation, *J. Manuf. Process.* 73 (2022) 112–127.

[34] S. Boopathi, An extensive review on sustainable developments of dry and near-dry electrical discharge machining processes, *J. Manuf. Sci. Eng.* 144 (2022) 1–18.

[35] P. Li, X. Li, F. Li, A novel recycling and reuse method of iron scraps from machining process, *J. Clean. Prod.* 266 (2020) 121732.

[36] X. Wu, C. Li, Z. Zhou, X. Nie, Y. Chen, Y. Zhang, H. Cao, B. Liu, N. Zhang, Z. Said, Circulating purification of cutting fluid: An overview, *Int. J. Adv. Manuf. Technol.* 117 (2021) 2565–2600.

[37] L. Simon, C.A.M. Moraes, R.C.E. Modolo, M. Vargas, D. Calheiro, F.A. Brehm, Recycling of contaminated metallic chip based on eco-efficiency and eco-effectiveness approaches, *J. Clean. Prod.* 153 (2017) 417–424.

[38] D.O.A. Carvalho, L.R.R. da Silva, L. Sopchenski, M.J. Jackson, Á.R. Machado, Performance evaluation of vegetable-based cutting fluids in turning of AISI 1050 steel, *Int. J. Adv. Manuf. Technol.* 103 (2019) 1603–1619.

2 Hybrid Advanced Machining for Sustainability in Manufacturing

Şenol Şirin

2.1 INTRODUCTION

Machining is one of the important manufacturing processes that changes the shape, dimensions, surface, and geometry of a workpiece by removing material in the form of chips (Kuntoğlu et al. 2020; Leksycki et al. 2021; Şirin et al. 2021). Especially after the industrial revolution, it has been understood that the production of technological products could only be possible with machining processes (Hofmann and Rüsch 2017; Oztemel and Gursev 2020). In other words, technological applications must be manufacturable with different machining processes. Machining processes can be examined under two groups (Figure 2.1). The first group includes conventional machining methods used to cut the workpiece with a cutting tool. Basic machining operations contain different processes, i.e., milling, turning, surface finishing process (grinding), drilling, etc. The second group is modern (nonconventional) machining such as electro chemical machining (ECM), laser machining (LBM), ultrasonic vibration machining (USM), abrasive water jet machining (AWJM), electric discharge machining (EDM), etc. Combining these techniques and adding some external energy sources like vibration, heat, and magnetic field etc., can develop variety of hybrid machining techniques (Lauwers et al. 2014). Hybrid machining process is defined as a process where two or more machining processes are combined to exploit their benefits simultaneously (Li et al. 2018a, 2018b; Liu et al. 2020). That hybridization potentially enhances the process efficiency, helps to reduce the cost, and prevents quality deterioration (Gupta et al. 2016). Hybrid advanced machining can be divided into two categories: (1) the combination of different energy sources moving simultaneously in the machining area (i.e., laser-assisted machining); (2) machining that combines two or more processes. Some important hybrid advanced machining techniques are laser-assisted machining, vibration-assisted machining, heat-assisted machining, electro chemical machining, and hybrid water jet machining.

Laser-Assisted Machining (LAM) is widely used for advanced high-strength machining (Zhang et al. 2022). LAM is becoming more and more popular in advanced hybrid machining due to its high laser beam intensity at low mean beam power,

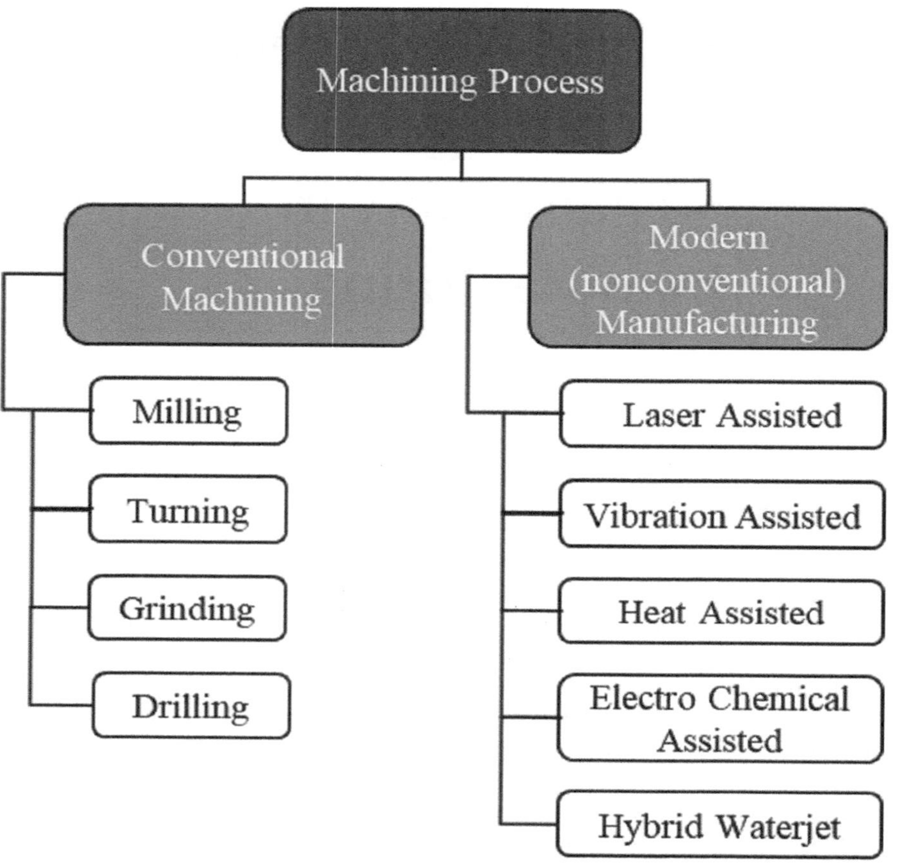

FIGURE 2.1 Important types of conventional and modern machining techniques and sources of assistance to develop hybrid machining techniques

good focusing characteristics because of the short pulse period, slight notch widths and limited heat-affected zones, productivity increases, low costs of production, and improved surface quality (Bean et al. 2018; Lakhdar et al. 2021; Krimpenis and Noeas 2022). In the LAM process, the movement of the laser softens the workpiece material, making it more accessible to machine high alloy steels or some ceramics (Vignesh and Ramanujam 2020). In this hybrid machining process, the laser beam focuses directly on the cutting tool, resulting in more effortless machining efficiency and higher machining performance (Li et al. 2022).

To enhance the machining performance, vibration-assisted machining (VAM) blends precision machining with small-amplitude tool vibration (Yang et al. 2020; Sharma et al. 2022). There are several processes, from milling to grinding to drilling in VAM. Significantly, vibration-assisted grinding is a new technology where combining conventional grinding and a vibration (usually in the ultrasonic range) is established (Gao et al. 2020; Yang et al. 2020). The tool is driven by reciprocating or elliptical motion in the VAM process. Thinner chips can be produced and cutting forces can be lowered. Consequently, according to conventional machining processes, VAM processes improve the surface quality and occur minimal burr (Greco et al. 2021).

Heat-Assisted Machining (HAM) has become a popular alternative to traditional machining methods for improving the machinability of hard-to-cut materials (Bijanzad et al. 2022). This HAM machining method has been researched since the latter part of the 19th century. Heat-Assisted Machining increases the difference in hardness of the cutting tool and workpiece by softening it, reducing the cutting forces, improving surface quality, and prolong the tool life (Varghese et al. 2019; Mruthunjaya and Yogesha 2021). Metals deform more rapidly when heated, improving their machinability (Salur et al. 2021) because heat generation helps the plastic deformation of difficult-to-machine materials.

For machinability enhancement and precise machining of metal-based materials, electro chemical machining (ECM) is also used in hybridization with turning, drilling, milling, and grinding (Ajith Arul Daniel et al. 2020). The plain ECM is preferred for mass production and is used to manufacture rigid or difficult-to-machine materials with conventional methods (Das et al. 2019; Chen et al. 2020). The ECM technique offers various benefits, including machining performance that is independent of the workpiece's material's mechanical characteristics, fabrication of a surface that is mostly stress-free, good surface quality and integrity, and improved productivity due to a higher material removal rate (MRR). In the metal sector, ECM provides a superior alternative—and occasionally the only one—to producing 3D complex-shaped features of materials or their components that are hard to manufacture (Kumar et al. 2018). Because of its operational flexibility and superior surface quality, hybrid water jet (HWJ) machining has drawn the attention of researchers and the industrial sector among the different modern hybrid machining processes (Pahuja et al. 2019). HWJ has been used extensively in the manufacturing industries to manufacture various materials such as metals and non-metals (ceramics, glass, etc.) (Kavya et al. 2020; Natarajan et al. 2020). When handling with materials that are hard to machine, such as metals, non-metals, composites, and ceramics, the HWJ machining technique produces so little heat in the cutting zone. Moreover, laser machining claims that HWJ offers superior surface quality and integrity than Wire EDM because of its greater material removal rate (Natarajan et al. 2020; Kanake and Ahuja 2022). When evaluated in general, Hybrid Advanced Machining processes have superior characteristics compared to their individual techniques (Wilkinson et al. 2019). In addition, in all hybrid advanced machining methods, it is seen that especially difficult-to-machine materials are easily shaped, cut, and machined.

2.2 DETAILS ON HYBRID ADVANCED MACHINING (HAM)

Hybrid machining combines one or more machining methods (Şirin 2022). Combining many unconventional methods provides the basis of the hybrid advanced machining process. Hybrid Advanced Machining processes can be listed as follows: laser-assisted machining (LAM), vibration-assisted machining (VAM), heat-assisted machining (HAM), electro chemical machining (ECM), and hybrid water jet machining (HWJ). The hybrid advanced machining processes listed in this book chapter were investigated, and the information gained was discussed in the following sections.

2.2.1 Laser-Assisted Machining (LAM)

After the advent of laser used in the 1970s, laser-assisted manufacturing (LAM) processes have found widespread use, such as powder metallurgy, forming, machining, and surface treatment (Anderson et al. 2006). High-strength materials, including aircraft, automobiles, energy, healthcare, and mining, are chosen for applications. These materials, however, are challenging to machine using standard methods. Short tool life, moderate cutting speeds, and frequent tool changes are detrimental to conventional machining procedures for difficult-to-machine materials (Anderson et al. 2006; Zhang et al. 2022). One of the methods of improving the machining properties of difficult-to-machine materials can be said to benefit from the thermal softening feature via heating the fabric with a heat source during

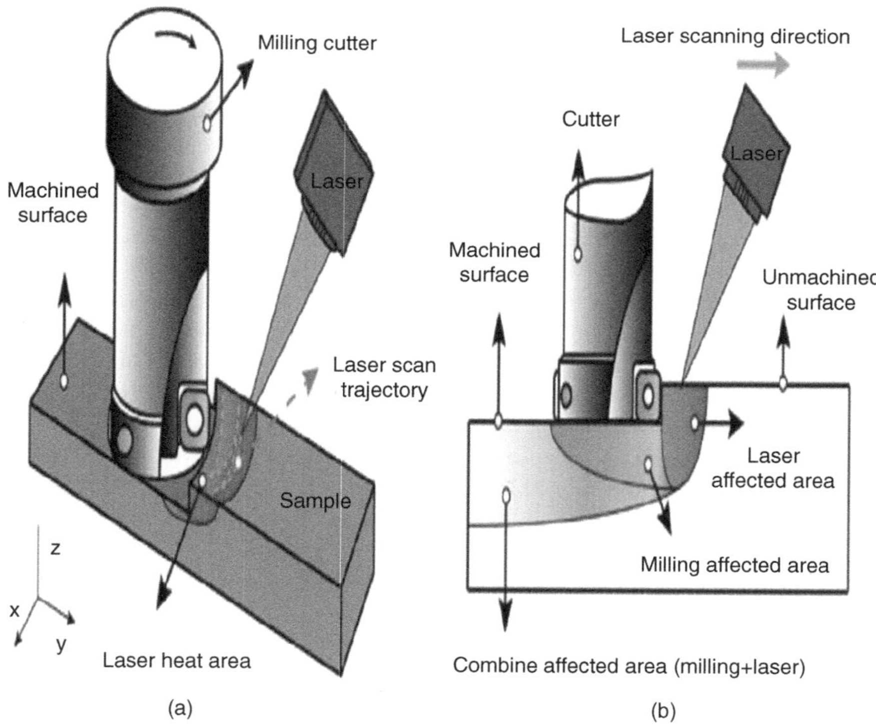

FIGURE 2.2 Heating effect of laser-assisted milling (LAMill) process, (a) Laser scanning trajectory for uniformly and (b) 3D heating the workpiece and heat-influenced superficial layer (Xu et al. 2020)

machining. Laser-Assisted Machining (LAM) has become a practical industrial method for processing materials that are challenging to work with. In the LAM machining process, the workpiece is heated with a concentrated laser beam before being cut using traditional cutting tools. The workpiece in front of the cutting zone receives highly localized heating from a laser during the LAM procedure. By lowering the material strength in a high-temperature cutting region, the LAM process provides a lower cutting force, slower tool wear rate, greater material removal rate, and improved surface quality (Zhang et al. 2022). The schematic representation of the LAM process is given in Figure 2.2.

Generally, the LAM process was preferred for conventional operations such as turning or milling (Anderson et al. 2006; Shang et al. 2019). For the efficiency of the LAM process, a single laser point is usually placed in front of the cutting tool as the heat source. Moreover, heating the workpiece homogeneously is the most critical

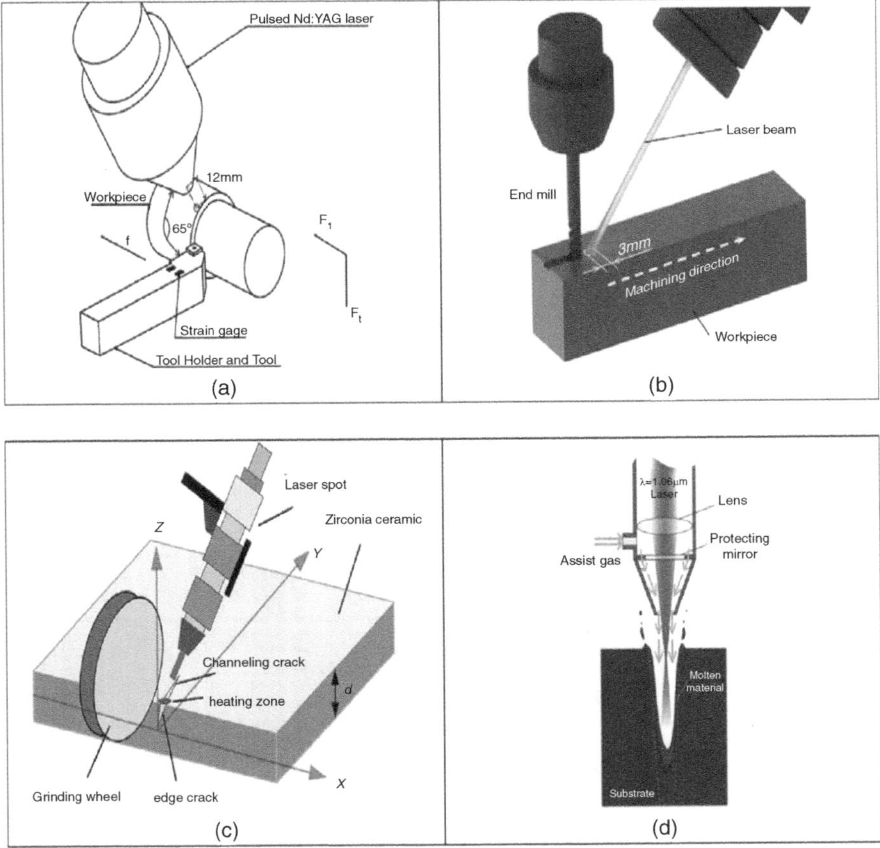

FIGURE 2.3 LAM process combined with traditional machining: (a) Turning (Abedinzadeh et al. 2022), (b) Milling (Cao et al. 2020), (c) Grinding (Ma et al. 2020), and (d) Drilling (Duan et al. 2020)

parameter to be considered in the LAM process (Zhang et al. 2022). Recently, to solve the homogeneous heating problem, a heat insertion control method has been proposed in the literature, ensuring that the workpiece is heated homogeneously in front of the cutting edge, especially in the milling process (Shang et al. 2019). LAM technology offers significant advantages compared to conventional machining, i.e., less power use during machining, higher material removal rates, reduction of cutting forces, reduction of chatter, minimization of residual stresses, reduction of tool wear, and improved surface quality. In addition, it has been stated in some studies that the overall processing cost can be reduced by up to 60% when LAM is used (Anderson et al. 2006). LAM-assisted traditional turning, milling, grinding, and drilling processes are given in Figure 2.3.

2.2.2 Vibration-Assisted Machining (VAM)

To increase machinability, the workpiece or cutting tool is subjected to high-frequency, low-amplitude mechanical energy vibrations during the vibration-assisted machining (VAM) process. Moreover, the VAM method can be applied to conventional machinings such as milling, turning, grinding, and drilling. Nonconventional processing technology VAM has started to be preferred in different areas, such as the application of aerospace parts, optics, biomedical science, and additive manufacturing technologies. Furthermore, the VAM method may be used to create superalloy materials like titanium and cobalt, tungsten, and nickel-based, as well as soft materials like copper and aluminum, difficult-to-machine materials

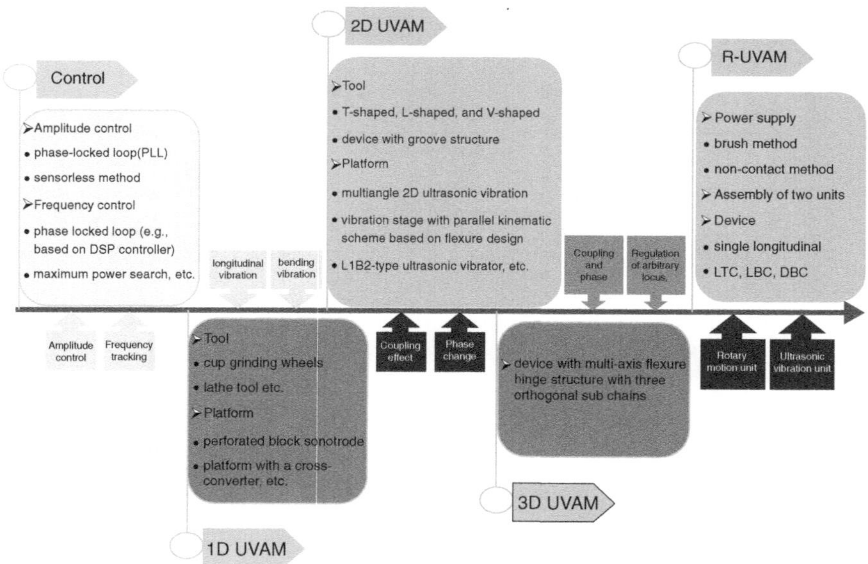

FIGURE 2.4 VAM process devices and applications. (Yang et al. 2020)

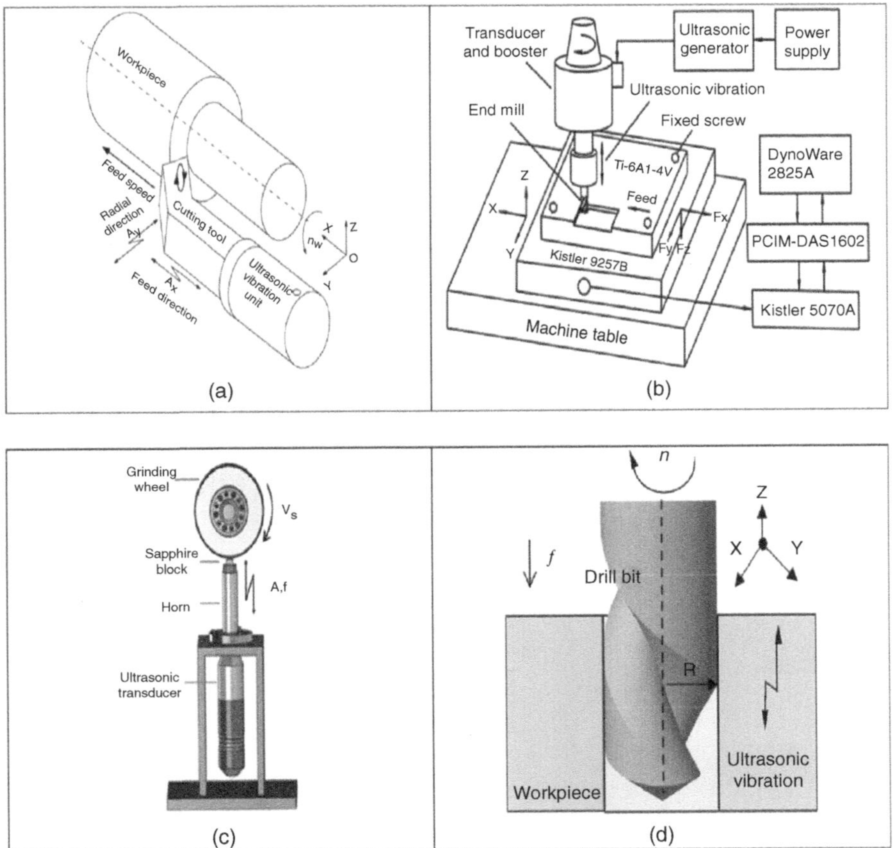

FIGURE 2.5 VAM combined with (a) Turning process (Wang et al. 2016), (b) Milling process (Gao et al. 2022), (c) Grinding process (Chen et al. 2022), and (d) Drilling process (Sorgato et al. 2021)

like high carbon and alloy steels, and soft materials like copper and aluminum. Table vibration and spindle vibration are the two different foundation operating principles VAM devices use (Yang et al. 2020). The first operating principles of the VAM process are 1-D (first used in the late 1950s), 2-D (presented in the 1990s), and 3-D (begun used in recent years). The second operating principle of the VAM process includes a rotary spindle system. The VAM operating procedures are given in Figure 2.4.

The VAM nontraditional process can be easily used in turning, milling, grinding, and drilling operations. It has started to be preferred in some studies in recent years, as it increases surface quality and reduces power consumption. Turning, milling, grinding, and drilling operations using the VAM process are given in Figure 2.5.

2.2.3 Heat-Assisted Machining (HAM)

HAM is a hybrid process used especially in machining materials such as superalloys. This nontraditional method, which made its name in the late 19th century, based on the easy machining of the workpiece heated by a source. The HAM process was employed in common applications like turning and milling at the end of the 20th century. Various heat sources, including laser, plasma, electron, and induction heat sources heat the workpiece. The life of the cutting tool and the surface quality of the workpiece both increase as a consequence of the machining of the cutting tool from the heated workpiece in the HAM process. Moreover, there has some advantages thanks to HAM, for example, power consumption and cutting tool wear were reduced, and workpiece finishing surfaces were improved. In the gas-based HAM process, the workpiece was heated using oxygen and liquid petroleum gas (LPG) together. Despite its ease of application, the gas-based HAM process had disadvantages such as low penetration, difficulty in controlling,

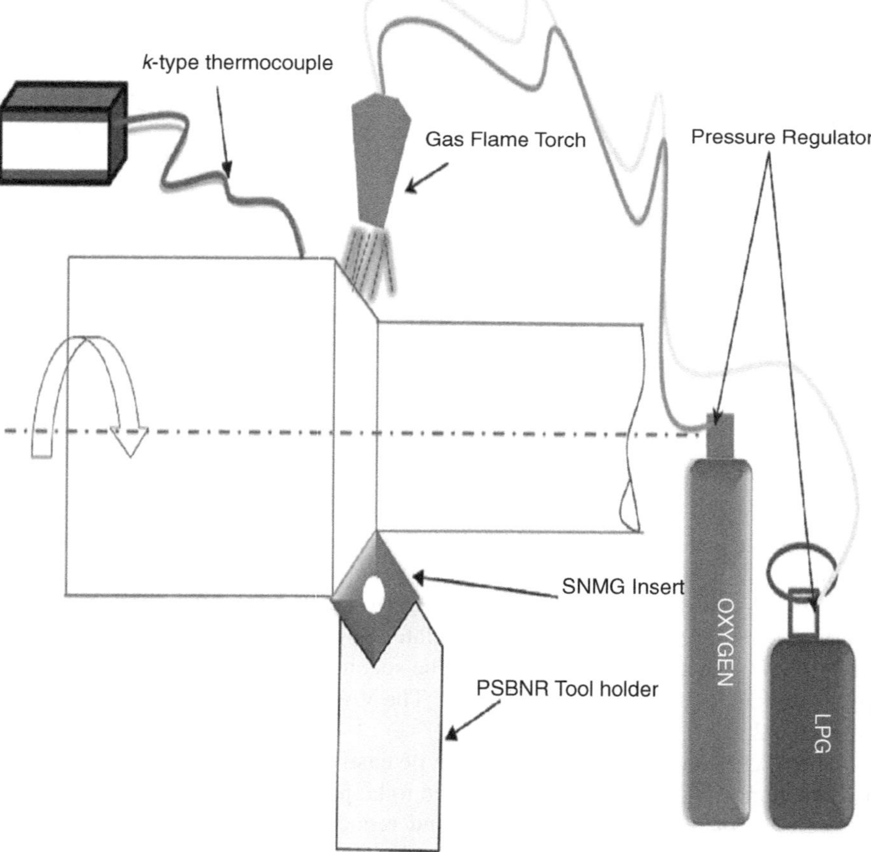

FIGURE 2.6   Gas-based HAM process details (Parida and Maity 2021)

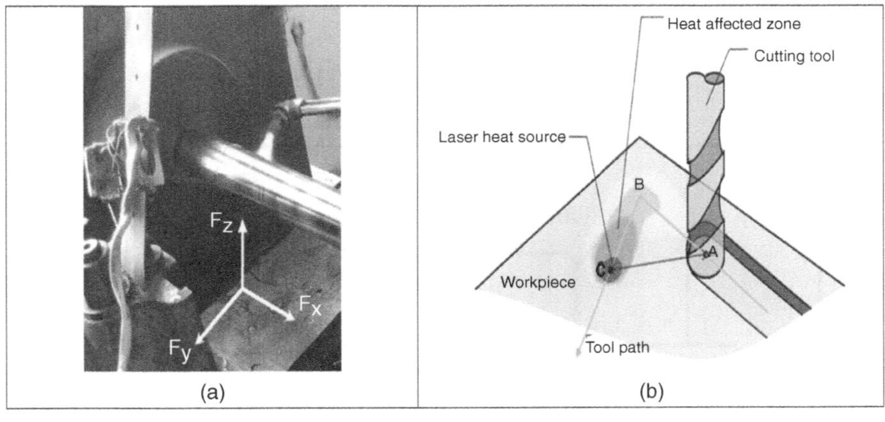

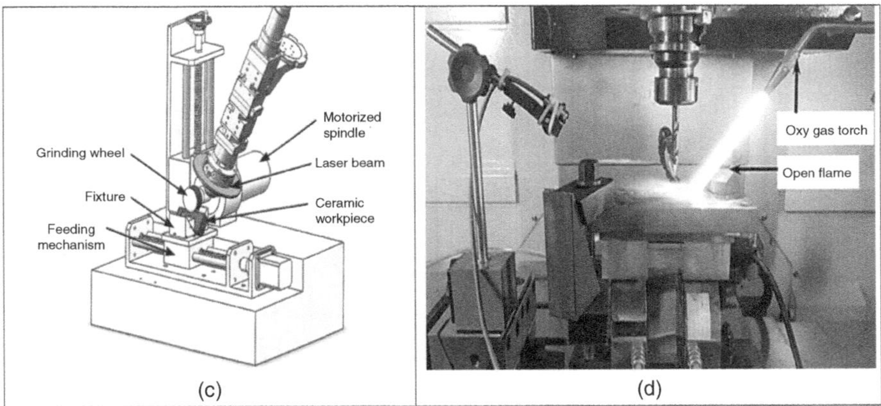

FIGURE 2.7   HAM process-assisted (a) Turning (Parida and Maity 2019), (b) Milling (Oh and Lee 2019), (c) Grinding (Ma et al. 2021), and (d) Drilling (Karabulut et al. 2022)

and oxidation in superalloy materials. The gas-based HAM process is given in Figure 2.6.

HAM process-assisted turning, milling, grinding, and drilling operations are given in Figure 2.7.

2.2.4 ELECTRO CHEMICAL MACHINING (ECM)

Electro Chemical Machining (ECM) process first appeared in Russia in 1928 (Bannard 1977). In recent years, the aviation sector has extensively used the ECM approach, particularly for producing intricate parts (Xu and Wang 2021). The ECM technology presents advantages such as no tool wear, high machining efficiency, lower machining costs compared to conventional machining, difficult-to-machine materials are easy to manufacture, and complex geometry products can be easily

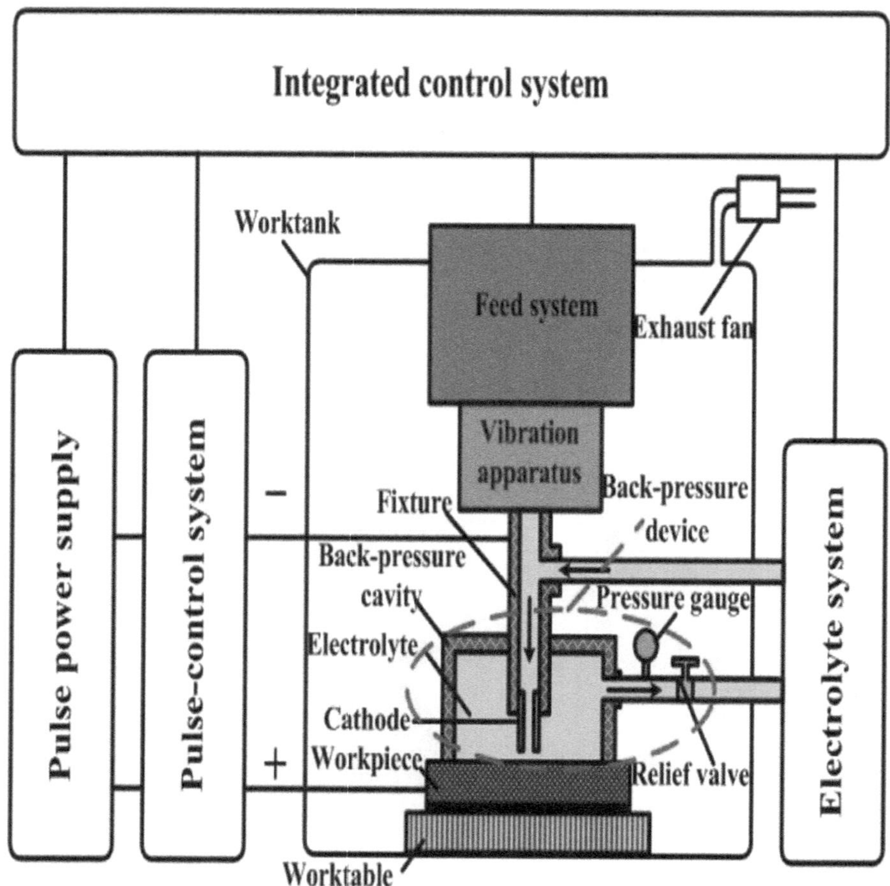

FIGURE 2.8 Machine setup and schematic representation in ECM process (Zhao et al. 2018)

obtained (Sous et al. 2022). The ECM process has a negatively charged cathode between the electrode and the workpiece material, conductive electrolytic fluid, etc., which is comparable to the electrical discharge machining (EDM) process. However, the main difference between the ECM and the EDM method is that no tool wear and sparks are created in the ECM method. The ECM process is between the electrically conductive tool (usually copper material -cathode) and the workpiece materials (+cathode). The chip forms in the ECM process are removed with the electrolysis liquid. The schematic representation of the ECM process was given in Figure 2.8.

The ECM process can be combined with conventional methods. Hybrid ECM-assisted methods to include turning, milling, grinding, and drilling traditional processes. These methods are given in Figure 2.9 respectively.

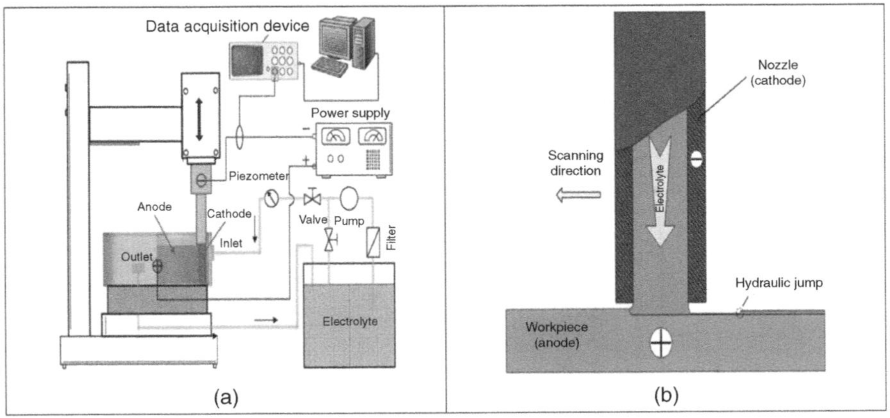

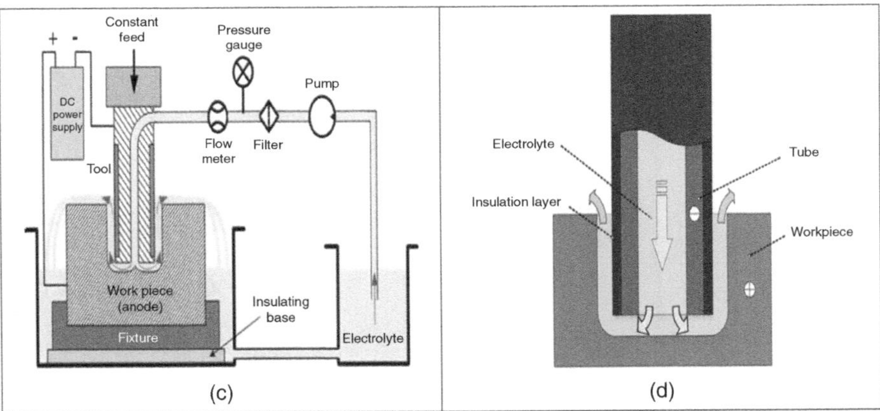

FIGURE 2.9   ECM-assisted (a) Turning (Ge et al. 2018), (b) Milling (Wang et al. 2019), (c) Grinding (Keerthivasan et al. 2021), and (d) Drilling (Zhang et al. 2019) processes

2.2.5 Hybrid Water Jet Machining (HWJ)

Due to its superior properties, water jet machining is widely preferred among advanced manufacturing processes in many industrial applications (Natarajan et al. 2020). Water jet machining (WJM) has some superior capabilities that can be listed, (1) excellent quality of the workpiece after the operation, (2) machining of fragile materials, and (3) it could cut almost entire materials for example, difficult-to-machine materials, superalloys, ceramic-based, metal-based, non-metals, composites, etc. The WJM process is divided into two different jet machining methods: abrasive water injection and abrasive suspension (Figure 2.10).

Hybrid water jet machining (HWJM) process can be combined with traditional turning, milling, grinding, and drilling operations (Figure 2.11).

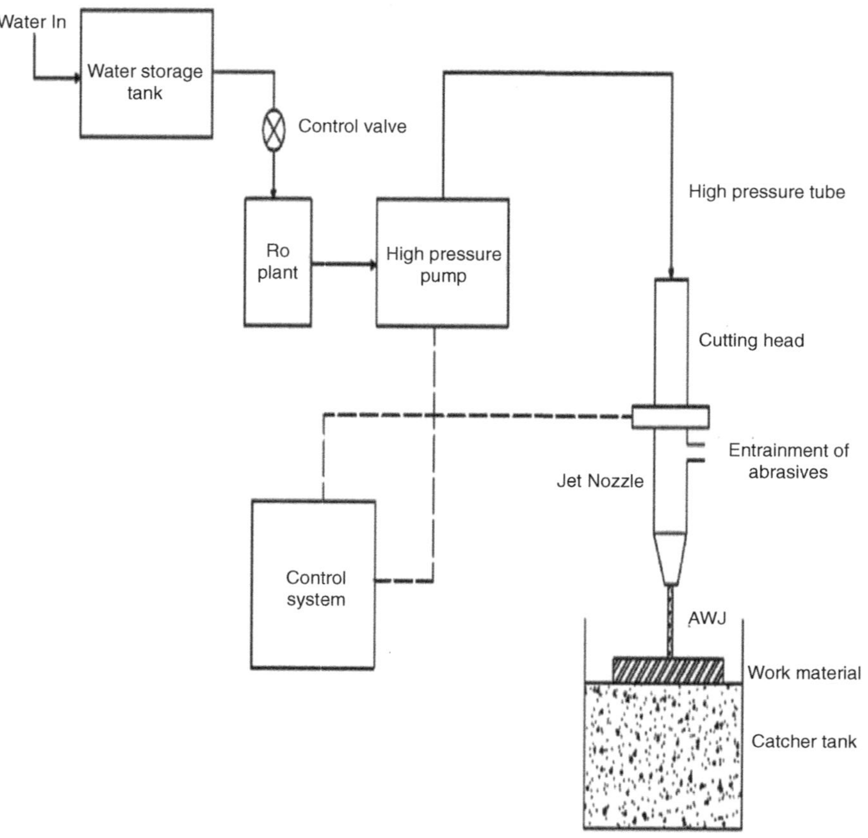

FIGURE 2.10   Experimental details and setup of water jet machining processes (Natarajan et al. 2020)

2.3 SUMMARY

This book chapter has highlighted the importance of a wide range of hybrid machining processes. A brief discussion on the application, working principle, and characteristics has also been done. The following important points can be mentioned to summaries:

- Using laser-assisted machining (LAM), high alloy steel and some ceramic material types can be machined more efficiently since the workpiece material gets softer via a laser source, especially in conventional machining such as turning, milling, grinding, and drilling.
- Vibration-assisted machining (VAM) offers some advantages such as reduced cutting forces, increased cutting tool life, improved surface finish, and decreased burr formation when combined with conventional techniques.

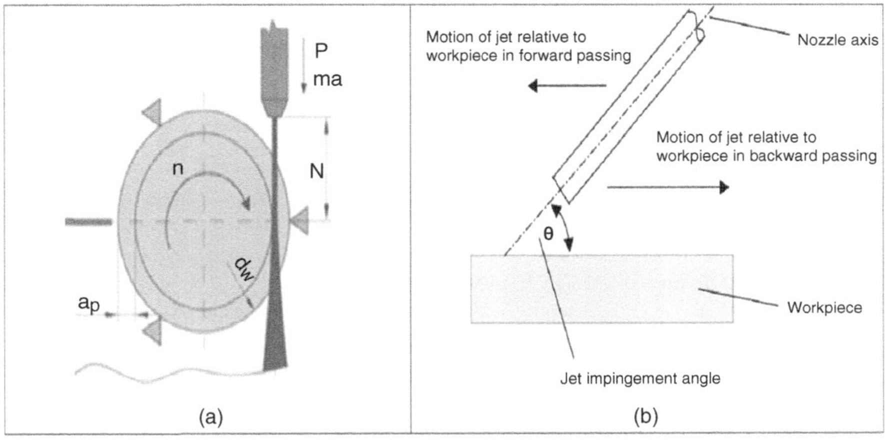

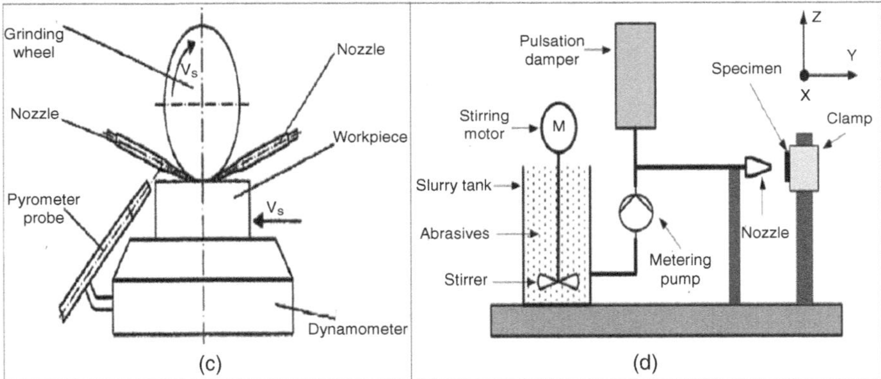

FIGURE 2.11 Hybrid water jet machining processes combined with traditional, (a) turning (Srivastava et al. 2019), (b) milling (Fowler et al. 2005), (c) grinding (Babic et al. 2005), and (d) drilling processes (Gao et al. 2018)

- In heat-assisted machining (HAM), the material is heated using an external heat source such as gas flames, coils etc. Enhanced machining performance can be obtained specially when machining hard materials.
- Hybridization of electro chemical machining (ECM) with conventional techniques can precisely cut a wide range of electrically conducting materials and obtain the required shape and size with tight tolerance.
- Hybrid water jet machining (HWJM), where water jet is the presence of abrasives is hybridized with turning, milling, and drilling, is an efficient process on materials with low machinability (such as superalloys, ceramics, glass, composites, etc.). This hybrid machining has made possible cutting of a variety of geometry parts with less machining stresses and damages to the work material.

In essence, it can be said that the HAM processes are superior to the individual machining processes (whether conventional or advanced types) and can still be explored more to establish the field further.

REFERENCES

Abedinzadeh, Reza, Ehsan Norouzi, and Davood Toghraie. 2022. "Study on Machining Characteristics of SiC–Al2O3 Reinforced Aluminum Hybrid Nanocomposite in Conventional and Laser-Assisted Turning." *Ceramics International* 48 (19). Elsevier: 29205–29216. doi:10.1016/J.CERAMINT.2022.05.196.

Ajith Arul Daniel, S., S. Vijay Ananth, A. Parthiban, and S. Sivaganesan. 2020. "Optimization of Machining Parameters in Electro Chemical Machining of Al5059/SiC/MoS2 Composites Using Taguchi Method." *Materials Today: Proceedings* 21 (January). Elsevier: 738–743. doi:10.1016/J.MATPR.2019.06.750.

Anderson, Mark, Rahul Patwa, and Yung C. Shin. 2006. "Laser-Assisted Machining of Inconel 718 with an Economic Analysis." *International Journal of Machine Tools and Manufacture* 46 (14). Pergamon: 1879–1891. doi:10.1016/J.IJMACHTOOLS.2005.11.005.

Babic, Deian, Darina B. Murray, and Andrew A. Torrance. 2005. "Mist Jet Cooling of Grinding Processes." *International Journal of Machine Tools and Manufacture* 45 (10). Pergamon: 1171–1177. doi:10.1016/J.IJMACHTOOLS.2004.12.004.

Bannard, J. 1977. "Electrochemical Machining." *Journal of Applied Electrochemistry* 7 (1). Springer: 1–29. doi:10.1007/BF00615526.

Bean, Glenn E., David B. Witkin, Tait D. McLouth, Dhruv N. Patel, and Rafael J. Zaldivar. 2018. "Effect of Laser Focus Shift on Surface Quality and Density of Inconel 718 Parts Produced via Selective Laser Melting." *Additive Manufacturing* 22 (August). Elsevier: 207–215. doi:10.1016/J.ADDMA.2018.04.024.

Bijanzad, Armin, Talha Munir, and Farouk Abdulhamid. 2022. "Heat-Assisted Machining of Superalloys: A Review." *International Journal of Advanced Manufacturing Technology* 118 (11–12). Springer Science and Business Media Deutschland GmbH: 3531–3557. doi:10.1007/s00170-021-08059-2.

Cao, Xu Feng, Wan Sik Woo, and Choon Man Lee. 2020. "A Study on the Laser-Assisted Milling of 13–8 Stainless Steel for Optimal Machining." *Optics & Laser Technology* 132 (December). Elsevier: 106473. doi:10.1016/J.OPTLASTEC.2020.106473.

Chen, Ji Peng, Lin Gu, and Guo Jian He. 2020. "A Review on Conventional and Nonconventional Machining of SiC Particle-Reinforced Aluminium Matrix Composites." *Advances in Manufacturing* 8 (3). Shanghai University: 279–315. doi:10.1007/s40436-020-00313-2.

Chen, Yue, Zhongwei Hu, Yiqing Yu, Zhiyuan Lai, Jiegang Zhu, Xipeng Xu, and Qing Peng. 2022. "Processing and Machining Mechanism of Ultrasonic Vibration-Assisted Grinding on Sapphire." *Materials Science in Semiconductor Processing* 142 (May). Pergamon: 106470. doi:10.1016/J.MSSP.2022.106470.

Das, Manojit, Debadutta Mishra, and Trupti Ranjan Mahapatra. 2019. "Machinability of Metal Matrix Composites: A Review." *Materials Today: Proceedings* 18 (January). Elsevier: 5373–5381. doi:10.1016/J.MATPR.2019.07.564.

Duan, Wenqiang, Xuesong Mei, Zhengjie Fan, Jichang Li, Kedian Wang, and Yifei Zhang. 2020. "Electrochemical Corrosion Assisted Laser Drilling of Micro-Hole without Recast Layer." *Optik* 202 (February). Urban & Fischer: 163577. doi:10.1016/J.IJLEO.2019.163577.

Fowler, G., P. H. Shipway, and I. R. Pashby. 2005. "A Technical Note on Grit Embedment Following Abrasive Water-Jet Milling of a Titanium Alloy." *Journal of Materials Processing Technology* 159 (3). Elsevier: 356–368. doi:10.1016/J.JMATPROTEC.2004.05.024.

Gao, Changshui, Zhuang Liu, Kai Zhao, and Chao Guo. 2018. "Abrasive Water Jet Drilling of Ceramic Thermal Barrier Coatings." *Procedia CIRP* 68 (January). Elsevier: 517–522. doi:10.1016/J.PROCIR.2017.12.106.

Gao, Honghong, Baoji Ma, Yuanpeng Zhu, and Heng Yang. 2022. "Enhancement of Machinability and Surface Quality of Ti-6Al-4V by Longitudinal Ultrasonic Vibration-Assisted Milling under Dry Conditions." *Measurement* 187 (January). Elsevier: 110324. doi:10.1016/J.MEASUREMENT.2021.110324.

Gao, Teng, Xianpeng Zhang, Changhe Li, Yanbin Zhang, Min Yang, Dongzhou Jia, Heju Ji, et al. 2020. "Surface Morphology Evaluation of Multi-Angle 2D Ultrasonic Vibration Integrated with Nanofluid Minimum Quantity Lubrication Grinding." *Journal of Manufacturing Processes* 51 (March). Elsevier: 44–61. doi:10.1016/J.JMAPRO.2020.01.024.

Ge, Yong Cheng, Zengwei Zhu, Zhou Ma, and Dengyong Wang. 2018. "Large Allowance Electrochemical Turning of Revolving Parts Using a Universal Cylindrical Electrode." *Journal of Materials Processing Technology* 258 (August). Elsevier: 89–96. doi:10.1016/J.JMATPROTEC.2018.03.013.

Greco, Sebastian, Katja Klauer, Benjamin Kirsch, and Jan C. Aurich. 2021. "Vibration-Assisted Micro Milling of AISI 316L Produced by Laser-Based Powder Bed Fusion." *Journal of Manufacturing Processes* 71 (November). Elsevier: 298–305. doi:10.1016/J.JMAPRO.2021.09.020.

Gupta, Kapil, N. K. Jain, and R. F. Laubscher. 2016. *Hybrid Machining Processes-Perspective on Machining and Finiahing.* Springer International Publishing. ISBN 978-3-319-25920-8.

Hofmann, Erik, and Marco Rüsch. 2017. "Industry 4.0 and the Current Status as Well as Future Prospects on Logistics." *Computers in Industry* 89 (August). Elsevier: 23–34. doi:10.1016/J.COMPIND.2017.04.002.

Kanake, Vikas, and B. B. Ahuja. 2022. "Hybrid Machining Processes: An Overview." *Materials Today: Proceedings*, September. Elsevier. doi:10.1016/J.MATPR.2022.09.335.

Karabulut, Şener, Musa Bilgin, Halil Karakoç, Navneet Khanna, and Murat Sarıkaya. 2022. "A Study on the Influence of Thermally Assisted Novel Hybrid Methods on the Drilling Behavior of Ti6Al4V Alloy." *Tribology International* 175 (November). Elsevier: 107852. doi:10.1016/J.TRIBOINT.2022.107852.

Kavya, J. T., R. Keshavamurthy, G. S. Pradeep Kumar, G. Ugrasen, and S. Manjoth. 2020. "Prediction of Machining Characteristics of Abrasive Water Jet Machined Al7075-TiB2 In-Situ Composite." *Materials Today: Proceedings* 24 (January). Elsevier: 851–858. doi:10.1016/J.MATPR.2020.04.395.

Keerthivasan, T., P. Kaushik, P. Mirunalini, and J. Surendhiran. 2021. "Design and analysis of erosion in electrochemical machining tool." *Materials Today: Proceedings* 37. Elsevier: 182–186.

Krimpenis, Agathoklis A., and Georgios D. Noeas. 2022. "Application of Hybrid Manufacturing Processes in Microfabrication." *Journal of Manufacturing Processes* 80 (August). Elsevier: 328–346. doi:10.1016/J.JMAPRO.2022.06.009.

Kumar, Pravin, Pradeep Jadhav, Mahavir Beldar, D. B. Jadhav, and Abhay Sawant. 2018. "Review Paper on ECM, PECM and Ultrasonic Assisted PECM." *Materials Today: Proceedings* 5 (2). Elsevier: 6381–6390. doi:10.1016/J.MATPR.2017.12.249.

Kuntoğlu, Mustafa, Abdullah Aslan, Danil Yurievich Pimenov, Üsame Ali Usca, Emin Salur, Munish Kumar Gupta, Tadeusz Mikolajczyk, Khaled Giasin, Wojciech Kapłonek, and Shubham Sharma. 2020. "A Review of Indirect Tool Condition Monitoring Systems and Decision-Making Methods in Turning: Critical Analysis and Trends." *Sensors 2021* 21 (1). Multidisciplinary Digital Publishing Institute: 108. doi:10.3390/S21010108.

Lakhdar, Y., C. Tuck, J. Binner, A. Terry, and R. Goodridge. 2021. "Additive Manufacturing of Advanced Ceramic Materials." *Progress in Materials Science* 116 (February). Pergamon: 100736. doi:10.1016/J.PMATSCI.2020.100736.

Lauwers, Bert, Fritz Klocke, Andreas Klink, A. Erman Tekkaya, Reimund Neugebauer, and Don McIntosh. 2014. "Hybrid Processes in Manufacturing." *CIRP Annals* 63 (2). Elsevier: 561–583. doi:10.1016/J.CIRP.2014.05.003.

Leksycki, Kamil, Feldshtein Eugene, and Ociepa Michał. 2021. "On the Effect of the Side Flow of 316l Stainless Steel in the Finish Turning Process Under Dry Conditions." *Facta Universitatis, Series: Mechanical Engineering* 19 (2): 335–343. doi:10.22190/FUME191118019L.

Li, Hansong, Shuxing Fu, Qingliang Zhang, Shen Niu, and Ningsong Qu. 2018a. "Simulation and Experimental Investigation of Inner-Jet Electrochemical Grinding of GH4169 Alloy." *Chinese Journal of Aeronautics* 31 (3). Elsevier: 608–616. doi:10.1016/J.CJA.2017.08.014.

Li, Lin, Azadeh Haghighi, and Yiran Yang. 2018b. "A Novel 6-Axis Hybrid Additive-Subtractive Manufacturing Process: Design and Case Studies." *Journal of Manufacturing Processes* 33 (June). Elsevier: 150–160. doi:10.1016/J.JMAPRO.2018.05.008.

Li, Maojun, Haobo Han, Xiaoyang Jiang, and Xiaogeng Jiang. 2022. "A Feasibility Study on High-Efficient Laser Cutting of SiC Particles Reinforced Aluminum Matrix Composite Using Single-Pass Strategy." *Optik* 265 (September). Urban & Fischer: 169485. doi:10.1016/J.IJLEO.2022.169485.

Liu, Jikai, Yufan Zheng, Yongsheng Ma, Ahmed Qureshi, and Rafiq Ahmad. 2020. "A Topology Optimization Method for Hybrid Subtractive–Additive Remanufacturing." *International Journal of Precision Engineering and Manufacturing – Green Technology* 7 (5). Korean Society for Precision Engineering: 939–953. doi:10.1007/s40684-019-00075-8.

Ma, Zhelun, Zhao Wang, Xuezhi Wang, and Tianbiao Yu. 2020. "Effects of Laser-Assisted Grinding on Surface Integrity of Zirconia Ceramic." *Ceramics International* 46 (1). Elsevier: 921–929. doi:10.1016/J.CERAMINT.2019.09.051.

Ma, Z., Q. Wang, J. Dong, Z. Wang, and T. Yu. 2021. "Experimental Investigation and Numerical Analysis for Machinability of Alumina Ceramic by Laser-Assisted Grinding". *Precision Engineering* 72. Elsevier: 798–806.

Mruthunjaya, M., and K. B. Yogesha. 2021. "A Review on Conventional and Thermal Assisted Machining of Titanium Based Alloy." *Materials Today: Proceedings* 46 (January). Elsevier: 8466–8472. doi:10.1016/J.MATPR.2021.03.490.

Natarajan, Yuvaraj, Pradeep Kumar Murugesan, Mugilvalavan Mohan, and Shakeel Ahmed Liyakath Ali Khan. 2020. "Abrasive Water Jet Machining Process: A State of Art of Review." *Journal of Manufacturing Processes* 49 (January). Elsevier: 271–322. doi:10.1016/J.JMAPRO.2019.11.030.

Oh, Won Jung, and Choon Man Lee. 2019. "A Study on Laser Assisted Acute Angle Milling Strategies and Preheating Distance." *Journal of Manufacturing Processes* 44 (August). Elsevier: 216–225. doi:10.1016/J.JMAPRO.2019.05.050.

Oztemel, Ercan, and Samet Gursev. 2020. "Literature Review of Industry 4.0 and Related Technologies." *Journal of Intelligent Manufacturing* 31 (1). Springer: 127–182. doi:10.1007/s10845-018-1433-8.

Pahuja, Rishi, M. Ramulu, and Mohamed Hashish. 2019. "Surface Quality and Kerf Width Prediction in Abrasive Water Jet Machining of Metal-Composite Stacks." *Composites Part B: Engineering* 175 (October). Elsevier: 107134. doi:10.1016/J.COMPOSITESB.2019.107134.

Parida, Asit Kumar, and Kalipada Maity. 2019. "Numerical and Experimental Analysis of Specific Cutting Energy in Hot Turning of Inconel 718." *Measurement* 133 (February). Elsevier: 361–369. doi:10.1016/J.MEASUREMENT.2018.10.033.

Parida, Asit Kumar, and Kalipada Maity. 2021. "Study of Machinability in Heat-Assisted Machining of Nickel-Base Alloy." *Measurement* 170 (January). Elsevier: 108682. doi:10.1016/J.MEASUREMENT.2020.108682.

Salur, Emin, Mustafa Acarer, and İlyas Şavkliyildiz. 2021. "Improving Mechanical Properties of Nano-Sized TiC Particle Reinforced AA7075 Al Alloy Composites Produced by Ball Milling and Hot Pressing." *Materials Today Communications* 27 (June). Elsevier: 102202. doi:10.1016/J.MTCOMM.2021.102202.

Shang, Zhendong, Zhirong Liao, Jon Ander Sarasua, John Billingham, and Dragos Axinte. 2019. "On Modelling of Laser Assisted Machining: Forward and Inverse Problems for Heat Placement Control." *International Journal of Machine Tools and Manufacture* 138 (March). Pergamon: 36–50. doi:10.1016/J.IJMACHTOOLS.2018.12.001.

Sharma, Ankit, Mohit Kalsia, Amrinder Singh Uppal, Atul Babbar, and Vikas Dhawan. 2022. "Machining of Hard and Brittle Materials: A Comprehensive Review." *Materials Today: Proceedings* 50 (January). Elsevier: 1048–1052. doi:10.1016/J.MATPR.2021.07.452.

Şirin, Emine, Turgay Kıvak, and Çağrı Vakkas Yıldırım. 2021. "Effects of Mono/Hybrid Nanofluid Strategies and Surfactants on Machining Performance in the Drilling of Hastelloy X." *Tribology International* 157 (May). Elsevier: 106894. doi:10.1016/j.triboint.2021.106894.

Şirin, Şenol. 2022. "Investigation of the Performance of Cermet Tools in the Turning of Haynes 25 Superalloy under Gaseous N2 and Hybrid Nanofluid Cutting Environments." *Journal of Manufacturing Processes* 76 (April). Elsevier: 428–443. doi:10.1016/J.JMAPRO.2022.02.029.

Sorgato, Marco, Rachele Bertolini, Andrea Ghiotti, and Stefania Bruschi. 2021. "Tool Wear Analysis in High-Frequency Vibration-Assisted Drilling of Additive Manufactured Ti6Al4V Alloy." *Wear* 477 (July). Elsevier: 203814. doi:10.1016/J.WEAR.2021.203814.

Sous, F., L. Heidemanns, T. Herrig, A. Klink, and T. Bergs. 2022. "Experimental Analysis on the Accuracy of Two Dimensional Curved Cuts in Wire ECM." *Procedia CIRP* 113 (January). Elsevier: 398–403. doi:10.1016/J.PROCIR.2022.09.190.

Srivastava, Ashish Kumar, Akash Nag, Amit Rai Dixit, Jiri Scucka, Sergej Hloch, Dagmar Klichová, Petr Hlaváček, and Sandeep Tiwari. 2019. "Hardness Measurement of Surfaces on Hybrid Metal Matrix Composite Created by Turning Using an Abrasive Water Jet and WED." *Measurement* 131 (January). Elsevier: 628–639. doi:10.1016/J.MEASUREMENT.2018.09.026.

Varghese, Vinay, K. Akhil, M. R. Ramesh, and D. Chakradhar. 2019. "Investigation on the Performance of AlCrN and AlTiN Coated Cemented Carbide Inserts during End Milling of Maraging Steel under Dry, Wet and Cryogenic Environments." *Journal of Manufacturing Processes* 43 (July). Elsevier: 136–144. doi:10.1016/J.JMAPRO.2019.05.021.

Vignesh, M., and R. Ramanujam. 2020. "Laser-Assisted High Speed Machining of Inconel 718 Alloy." *High-Speed Machining*, January. Academic Press: 243–262. doi:10.1016/B978-0-12-815020-7.00009-6.

Wang, Qiang, Yongbo Wu, Jia Gu, Dong Lu, Yuebo Ji, and Mitsuyoshi Nomura. 2016. "Fundamental Machining Characteristics of the In-Base-Plane Ultrasonic Elliptical Vibration Assisted Turning of Inconel 718." *Procedia CIRP* 42 (January). Elsevier: 858–862. doi:10.1016/J.PROCIR.2016.03.008.

Wang, Xindi, Ningsong Qu, and Xiaolong Fang. 2019. "Reducing Stray Corrosion in Jet Electrochemical Milling by Adjusting the Jet Shape." *Journal of Materials Processing Technology* 264 (February). Elsevier: 240–248. doi:10.1016/J.JMATPROTEC.2018.09.017.

Wilkinson, N. J., M. A. A. Smith, R. W. Kay, and R. A. Harris. 2019. "A Review of Aerosol Jet Printing—A Non-Traditional Hybrid Process for Micro-Manufacturing." *The International Journal of Advanced Manufacturing Technology* 105 (11). Springer: 4599–4619. doi:10.1007/S00170-019-03438-2.

Xu, Dongdong, Zhirong Liao, Dragos Axinte, Jon Ander Sarasua, Rachid M'Saoubi, and Anders Wretland. 2020. "Investigation of Surface Integrity in Laser-Assisted Machining of Nickel Based Superalloy." *Materials & Design* 194 (September). Elsevier: 108851. doi:10.1016/J.MATDES.2020.108851.

Xu, Zhengyang, and Yudi Wang. 2021. "Electrochemical Machining of Complex Components of Aero-Engines: Developments, Trends, and Technological Advances." *Chinese Journal of Aeronautics* 34 (2). Elsevier: 28–53. doi:10.1016/J.CJA.2019.09.016.

Yang, Zhichao, Lida Zhu, Guixiang Zhang, Chenbing Ni, and Bin Lin. 2020. "Review of Ultrasonic Vibration-Assisted Machining in Advanced Materials." *International Journal of Machine Tools and Manufacture* 156 (September). Pergamon: 103594. doi:10.1016/J. IJMACHTOOLS.2020.103594.

Zhang, Hang, Rong Yan, Ben Deng, Jieyu Lin, Minghui Yang, and Fangyu Peng. 2022. "Investigation on Surface Integrity in Laser-Assisted Machining of Inconel 718 Based on In-Situ Observation." *Procedia CIRP* 108 (C). Elsevier: 129–134. doi:10.1016/J. PROCIR.2022.03.025.

Zhang, Yuhang, Ningsong Qu, Xiaolong Fang, and Xindi Wang. 2019. "Eliminating Spikes by Optimizing Machining Parameters in Electrochemical Drilling." *Journal of Manufacturing Processes* 37 (January). Elsevier: 488–495. doi:10.1016/J.JMAPRO. 2018.12.024.

Zhao, J., F. Wang, X. Zhang, Z. Yang, Y. Lv, and Y. He. 2018. "Experimental Research on Improving ECM Accuracy and Stability for Diamond-Hole Grilles." *Procedia CIRP*, 68. Elsevier: 684–689.

3 Wire EDM and Laser Beam Machining of NiTi Alloys for Overall Sustainability

M. Adam Khan, G. Ebenezer
and J. T. Winowlin Jappes

3.1 INTRODUCTION

Sustainability is a global concern these days. It is well known that environment, economy, and society are the three pillars of sustainability. To strengthen these three pillars, there are some key-drivers such as cost-efficiency, environment-friendliness, safety, and quality. These key-drivers can be attained by techniques such as green machining, process optimization, hybrid manufacturing, modern machining, sustainable cooling, and lubrication. Modern machining processes have been recognized as viable alternates of the conventional machining processes, especially in case of machining difficult-to-machine materials. Wire-EDM and laser beam machining are two of them. When any machining process helps to secure good quality at low cost and with low environmental footprints and involves safe operational practices, then it is said to achieve overall sustainability. NiTi alloy is a specific purpose material with a submicroscopic structure of 3D grid of each nickel atom surrounded by titanium atoms (Li et al. 2010). It is evident that the structure of NiTi alloy is body-centred cubic (BCC), i.e., nickel atom surrounded by titanium atoms. NiTi alloy's mechanical and electrical properties were taken from the literature and furnished in Table 3.1 (Behera et al. 2020; Choudhary et al. 2016; Naresh et al. 2016; Sharma et al. 2015; Tanzi et al. 2019; Wang et al. 2018).

As known, nickel and titanium are the major constituents of the NiTi alloy system, the structural properties of both nickel and titanium influence the properties of the NiTi alloy. The structural properties of both nickel and titanium are tabulated in Table 3.2 (Behera et al. 2020; Choudhary et al. 2016).

NiTi alloy is a super elastic material capable of repossession and withstanding huge strain (Hassan et al. 2014). The other prominent properties of the NiTi alloy are high strength-to-mass ratio, better toughness, superior shape memory effect (SME), excellent biocompatibility, improved pseudo-elasticity, high resistance to wear, and corrosion (Andani et al. 2017; Choudhary et al. 2016; Hassan et al. 2014; Manjaiah

TABLE 3.1

Properties of Nickel Titanium (NiTi) alloy

Properties	Ni-Ti alloy	Units
Modulus of elasticity	83	GPa
Yield strength	195–690	MPa
Ultimate strength	895	MPa
Melting temperature	1,300	°C
Density	6.45	$\dfrac{\text{g}}{\text{cm}^3}$
Thermal conductivity	18	$\dfrac{\text{W}}{\text{cm}°\text{C}}$
Recovered elongation	8	%
Coefficient of thermal expansion	11.0×10^{-6}	$\dfrac{\text{mm}}{\text{mK}}$

TABLE 3.2

Structural properties of the Ni and Ti

Properties	Ni	Ti	Unit
Nuclear charge number	28	22	–
Average mass of an atom	58.6934	47.867	amu
Melting point	1,453	1,660	°C
Boiling point	2,732	3,287	°C
Density (30°C)	8.908	4.506	$\dfrac{\text{g}}{\text{cm}^3}$
Heat of fusion	17.48	14.15	$\dfrac{\text{kJ}}{\text{mol}}$
Enthalpy of vaporization	377.5	425	$\dfrac{\text{kJ}}{\text{mol}}$
Molar heat capacity	26.07	25.06	$\dfrac{\text{J}}{\text{molK}}$
Thermal conductivity	90.9	21.9	$\dfrac{\text{W}}{\text{mK}}$
Thermal expansion (25°C)	13.4	8.6	$\dfrac{\text{mm}}{\text{mK}}$
Magnetic ordering	Ferromagnetic	Paramagnetic	–

et al. 2014; Singh et al. 2021a, 2021b). Due to the astounding properties, the applications of the NiTi alloy were found global, some of them are revertible frames for spectacles, orthopaedical transplants, orthodontic clasps, guide wires in construction industries, filters in blood transfusion devices, surgical devices, cardiovascular stents, micro-electromechanical system (MEMS) devices, cellular phone antennas, automotive fasteners, micro sensors, micro actuators, high torque transmitting coupling, helicopter rotor shaft (Choudhary et al. 2016; Dash et al. 2019; Patel et al. 2020; Qadir et al. 2021).

The NiTi alloy had better physical and mechanical properties compared with stainless steel, tantalum, magnesium, iron. The equiatomic $Ni_{50}Ti_{50}$ alloy possessed appreciable elastic effect and microhardness. The microhardness of the $Ni_{50}Ti_{50}$ alloy was increased by 27% due to the heat treatment. The temperature of heat treatment is 865°C and the same maintained for 15 mins. After heat treatment, the microstructure of the $Ni_{50}Ti_{50}$ alloy was transformed into homogeneous structure (Mohammed et al. 2020).

The physical property, mechanical property, corrosion resistance, and the biocompatibility of the NiTi alloy was improved by applying coatings over the surface. Few common coatings in current scenario were diamond-like carbon (DLC), tantalum oxide, polyetheretherketone, etc. The same were improved by alloying specific element based on the applications. Some of the modified NiTi system were aluminium–nickel–titanium (Al–Ni–Ti), aluminium–cobalt–nickel–titanium (Al–Co–Ni–Ti), aluminium–ferrum–nickel–titanium (Al–Fe–Ni–Ti), nickel-titanium-copper (Ni–Ti–Cu), etc. In addition to that, the porosity in the NiTi alloy also plays a vital role in ensuring the biocompatibility of implant materials. The porous NiTi alloy was found to be best among the biocompatible materials. (Bahraminasab et al. 2011; Barras and Myers 2000; Hornbuckle et al. 2015; Khanlari et al. 2018; Minshull et al. 2011; Mohammed et al. 2020; Zhang et al. 2018).

The $Ni_{50}Ti_{50}$ alloy exists in austenitic phase and martensitic phase, in some cases it exists in rhombohedral phase. Mohammed et al. reported that in $Ni_{50}Ti_{50}$ alloy, both monoclinic and rhombohedral phases coexist i.e., martensitic and R phase. The X-ray diffraction pattern represents the coexistence of both monoclinic and rhombohedral phases. The reason for the coexistence was found to be as very low martensitic finish temperature. This low temperature was led to inadequate transformation of the $Ni_{50}Ti_{50}$ alloy (Mohammed et al. 2020). The $Ni_{50}Ti_{50}$ alloy has been extensively used in bioimplants, due to the property of anti-thrombosis and super elasticity (Barras and Myers 2000). It was also found by extensive experimentation the $Ni_{50}Ti_{50}$ alloy is capable of withstanding the high pressure and high strain, which made the $Ni_{50}Ti_{50}$ alloy employable as protective structures in space reentry vehicles (Zhang et al. 2018).

Some literature reported that very high Ni-rich NiTi alloys were unsuitable for most elastic applications. But in contradiction, $Ni_{53}Ti_{47}$, $Ni_{54}Ti_{46}$, $Ni_{55}Ti_{45}$, $Ni_{56}Ti_{44}$, $Ni_{57}Ti_{43}$, $Ni_{58}Ti_{42}$, alloys were with notable biocompatible properties and convincing hardness. Among the alloys, $Ni_{55}Ti_{45}$, $Ni_{56}Ti_{44}$, showed a maximum hardness of 644 ± 5.4 VHN (Vickers hardness number) ≈ 58 HRC (Rockwell hardness C scale). The shape memory capability of the NiTi alloy has varied with the variation in the weight percentage of Ni. The materials $Ni_{55}Ti_{45}$, and $Ni_{60}Ti_{40}$ alloy found communal usage in widespread applications. $Ni_{55}Ti_{45}$ alloy was reported with unique shape memory properties compared to other high Ni-rich NiTi alloys (Hornbuckle et al. 2015). The $Ni_{60}Ti_{40}$ alloy was found to be better than $Ni_{55}Ti_{45}$ alloy and other conventional materials. The alloy exhibits higher hardness ranges between 58 HRC and 62 HRC. On the other hand, low elastic modulus ($\approx$ 100 GPa), large strain recovery ($\approx$ 5%) and moderate density ($\approx$ 6.75 g/cm^3) are much lower than steel, so it can act as a viable replacement for bone transplant material (Bahraminasab et al. 2011; Elloy et al. 1976; Khanlari et al. 2018; Matassi et al. 2013).

The Titanium (Ti) rich NiTi alloy was used in dental braces, drug delivery systems, self-expanding stents, cardiovascular devices and ophthalmology. Some of the specific properties of Ti-rich NiTi alloy were reported in the literature: biocompatibility, corrosion resistance, magnetic resonance imaging (MRI) compatibility, and kink resistance (Paula et al. 2008; Tadayyon et al. 2016, 2017; Zhu et al. 2019).

3.2 NITI ALLOY MACHINING METHODS

Achieving the looked-for functionality and dimensional stability of machined products is a key challenge in machining of NiTi alloy. Specifically, NiTi alloys exhibit high resistance to cut and high mechanical strength made them difficult to perform machining. Machining such hard and difficult-to-cut NiTi alloy is a major task for any manufacturing industry. Additionally, during the machining processes it is supposed to maintain certain required properties of the NiTi alloy throughout the process. Some properties of NiTi alloy, such as more ductility and long elastic limit, resulted in severe burr formation, and cyclic hardening made the machining operation more difficult (Arunachalam and Mannan 2000; Ezugwu et al. 2005; Uthayakumar et al. 2016). The conventional machining processes are classified based on micro and macro machining. On the other hand, the non-conventional ways of machining processes are classified based on the energy utilized for the machining process. There are four forms of energy utilized for the machining process; they are mechanical, thermo-electrical, chemical, and electro chemical.

The mechanical form of machining uses the concept of erosion and mechanical ablation to cut the material from the workpiece surface. Since the cutting of workpiece material is made by means of slurry of liquids and particles, the surface finish is also determined by the speed of machining operation, aka "feed", velocity of the particle, aka "jet velocity", jet pressure, and size of the particle. The ultrasonic machining (USM), abrasive jet machining/abrasive jet blasting (AJM/AJB), water jet machining (WJM), and abrasive water jet machining (AWJM) are categorized under mechanical methods of non-traditional machining process. Abrasive particles-based machining is for hard materials and water and slurry-based machining for soft and moderately hard materials.

Thermal or thermo-electric energy machining processes use the heat energy to liquify and then to evaporate the material from the surface. By continuous melting and evaporation, the removal of surface metal takes place. The thermal energy has been generated by means of spark between electrode and workpiece in electrical discharge machining, amplified mono chromatic radiation in laser machining, and high energy electron beam in electron beam machining. In chemical and electro-chemical machining processes, the machining process occurs by means of chemical ablation and ion displacement respectively. The electrical discharge machining (EDM) has been identified as one of the most promising thermo-electrical machining methods for machining NiTi alloys. The EDM machining process is sub-classified into wire electrical discharge machining (WEDM), micro electrical discharge machining (μ EDM), rotary EDM, and dry EDM based on the requirement. Other convincible thermo-electric machining processes for micro machining are identified as laser beam machining (LBM), ion beam machining (IBM), electron beam machining (EBM), and plasma beam machining (PBM). The machining of NiTi alloys through thermal-based machining processes was found to be satisfied, but the problem lies with the generation of thermal cracks, more heat-affected zone due to high thermal energy required for melting and formation of white cast layer on the surface due to the redeposition of the molten material. The chemical machining (CHM) and electrochemical machining (ECHM) methods use chemical corrosion/reverse electroplating perception to eradicate unwanted material from the surface (Ezugwu et al. 2005; Goyal and Rahman 2021; Kulkarni et al. 2020; Muhammad et al. 2014; Soni et al. 2017; Uthayakumar et al. 2016). The detailed classification of the conventional and non-conventional machining processes is depicted in Figure 3.1.

Based on the literatures, conventional means of machining NiTi alloy was unsuccessful because of lower heat conduction property of this alloy. During the machining the heat was generated on the surface of the tool and workpiece due to the friction. Due to lower heat conduction of NiTi alloy, heat was intensified in-between the tool and work piece. The prolonged intensification of heat was resulted in the high tool wear, surface damage and deprived surface finish. To overcome the issues faced in the conventional machining process, a non-conventional machining process was employed. The non-conventional machining process are classified based on energy, viz., mechanical, thermal, chemical, and electro chemical. When it comes to mechanical energy-based machining process, say abrasive machining process hard particles erode the workpiece by abrasion (Arunachalam and Mannan 2000; Ezugwu et al. 2005; Mandal et al. 2014; Muhammad et al. 2014; Nishanth et al. 2019; Uthayakumar et al. 2016).

Furthermore, the thermal-based non-conventional machining processes, viz., electric discharge machining (EDM), laser beam machining (LBM), wire electric discharge machining (WEDM), ion-beam machining (IBM), plasma beam machining (PBM), etc., have less thermal conduction for the material especially with low thermal conductivity. Apart from this, heat-affected zone formed during the machining process also affects the performance of machining (Arunachalam and Mannan 2000; Mandal et al. 2014; Nishanth et al. 2019; Slătineanu et al. 2020).

3.3 WIRE ELECTRIC DISCHARGE MACHINING (WEDM)

Wire Electric Discharge Machining (WEDM) is a state-of-the-art machining technique employed to cut hard conductive materials. The tool electrode is typically a wire, whose size ranges between 0.10 and 0.30 mm. The wire feed mechanism is used to control the feed rate of electrode wire and maintain the wire tension throughout the process. The commonly used materials for tooling in WEDM process found in the literatures are namely, copper, graphite, copper tungsten, silver tungsten, copper

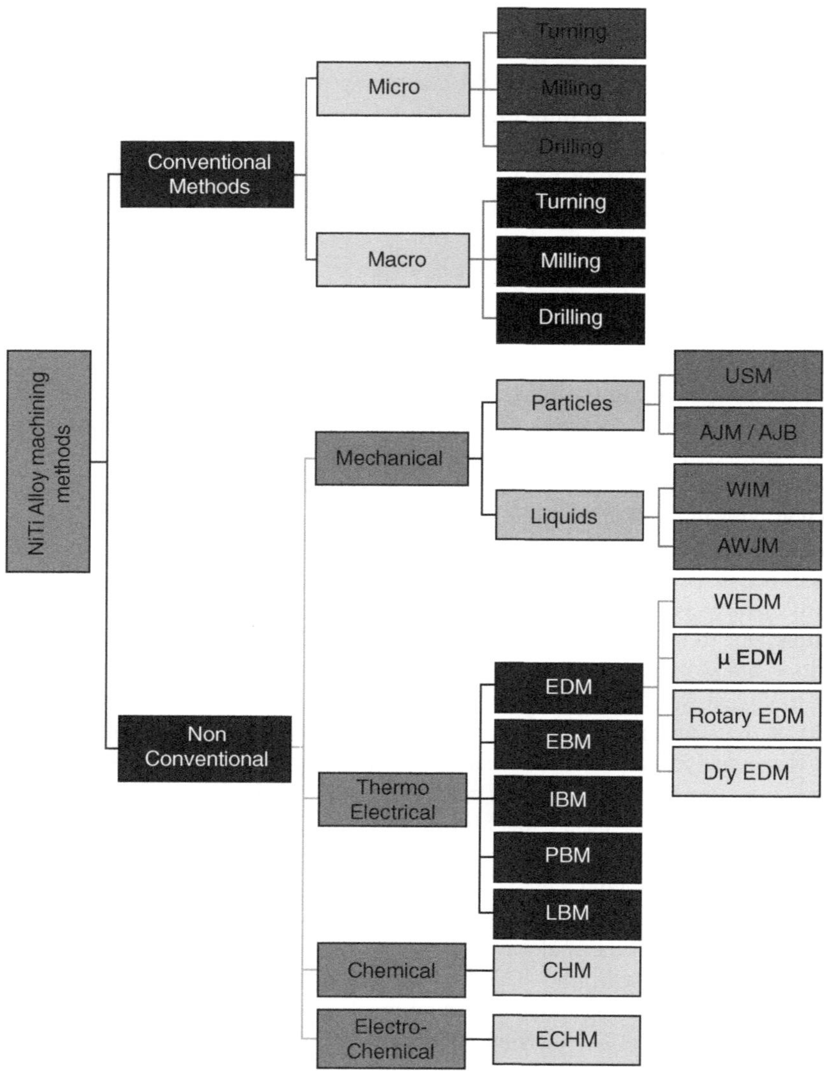

FIGURE 3.1   Classification of machining processes based on different source of energy

graphite, zinc, and brass (Jilani et al. 1984; Maher et al. 2015; Slătineanu et al. 2020; Valaki et al. 2016). A high voltage generator and a pulse control mechanism are available to control the high potential difference at continuous interval. The dielectric fluid is used as a medium to control flow of electricity and for flushing redeposited slag material over the machined surface after vaporization. The dielectric fluids found in commercial usage are, deionized water, kerosene, paraffin oil, blend of distilled water. Apart from the above-mentioned some researchers tried tap water (Jilani et al. 1984) and vegetable oil (Valaki et al. 2016) as dielectric fluids.

The machining initiated by providing high potential in-between the wire electrode and workpiece, immersed in the dielectric medium. Since the gap maintained in-between the electrode wire and the workpiece is low, an electrical spark is generated. The intensity of the spark depends on the applied voltage and the gap in-between the electrode wire and the workpiece. The spark is an indication of electricity conducted in-between the electrode wire and workpiece. The electric spark is capable of generating high thermal energy in form of temperature ranging from 8,000°C to 12,000°C (Abdulkareem et al. 2010). This spark repeated multiple times in a second, capable of eroding small surface material by means of melting and vaporization. After vaporization of material, partially melted and non-vaporized melted material flushed away by means of high-pressure dielectric fluid flowing in between electrode wire and workpiece.

The typical control parameters responsible for the machinability and the quality of the machined specimens were identified as spark ON time, spark OFF time,

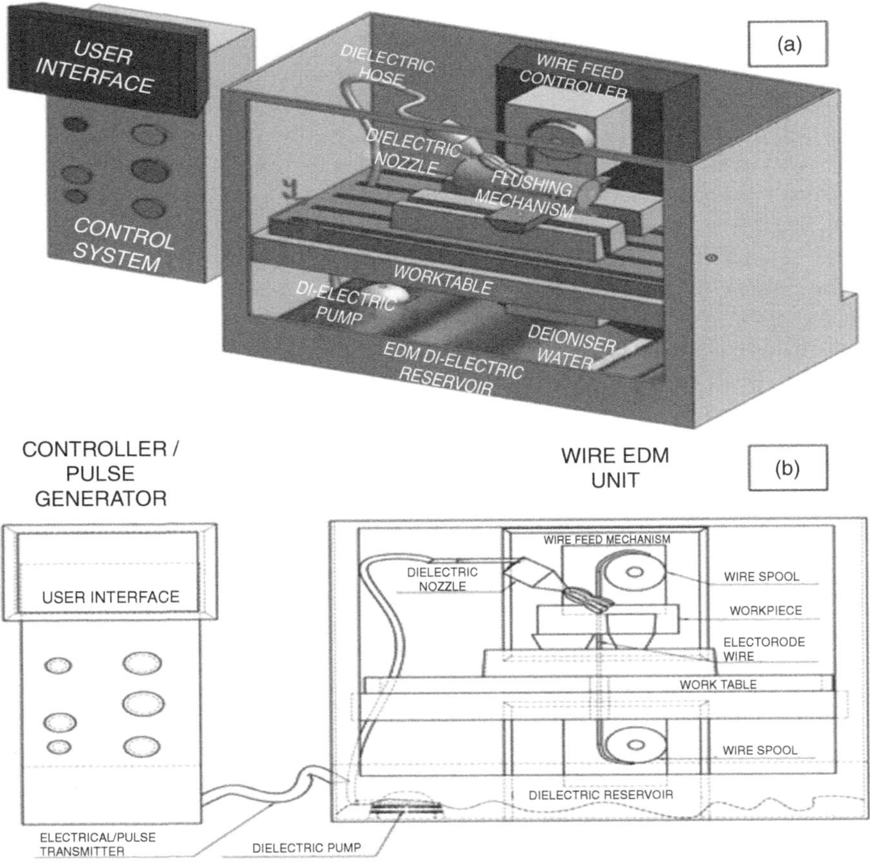

FIGURE 3.2 Basic arrangement of WEDM setup, (a) 3D view and (b) schematic view

applied voltage, peak current, wire feed rate, wire tension, dielectric flushing pressure, dielectric medium, wire material (Chaudhari et al. 2020; Daneshmand et al. 2013; Goyal and Rahman 2021; Kulkarni et al. 2017, 2020; Kumar et al. 2019; Soni et al. 2017). The basic schematic of WEDM machining is depicted in Figure 3.2.

Most of the literatures reported that the effect of spark ON (S_{ON}) time mostly influences the output characteristics of the WEDM machining technique. The most studied output characteristics were surface quality of the machined surface, optical characteristics of the machined surface, cutting efficiency of the machining process (Goyal and Rahman 2021; Kulkarni et al. 2017, 2020; Soni et al. 2017). Most of the researchers reported that the increase in the S_{ON} time increased the spark duration, which has enabled the sparking to last more. Higher sparking time has increased the commotion of melting, and this consecutively improved the material removal. Due to that, the S_{ON} time directly influences the surface quality and the cutting rate of the machining process (Chaudhari et al. 2020; Daneshmand et al. 2013; Singh et al. 2021).

Kulkarni et al. (2018) reported that during the machining of medical grade NiTi alloy using the WEDM process, S_{ON} time was identified as the noteworthy parameter. The roughness average (Ra) was found to be increased with the increase in the S_{ON} time. Like Ra, cutting rate was also increased with the increase in the S_{ON} time. The same results were reported by most of the literatures. Some of the researchers took peak current and the wire parameters such as wire material, wire feed rate, wire tension as input parameters. They concluded that peak current was the slightest influenced parameter on roughness average (Ra) compared with S_{ON} time (Gupta et al. 2021; Hodgson et al. 2021; Kanlayasiri and Boonmung 2007; Reddy et al. 2015; Singh et al. 2021).

In contrary, Doreswamy et al. reported that removal of material increased approximately 93% for the increase in peak current from 2 A to 6 A. The authors also confirmed that more surface defects like, cracks, voids were found for maximum peak current (Doreswamy et al. 2022). Hargovind Soni and his team (Soni et al. 2017) investigated the impact of the process parameters on the microstructure and micro hardness of the machined specimens. Surface defects such as, globules, craters, micro pores and recast layer were seen. The microstructure of the machined specimens revealed that the defects were found much nearer the machined surface. The microhardness was maximum at the machined surface and keeps on decreasing as the distance from the outer surface increases. Some of the literatures reported that the formation of oxide layer on the machined surface might be the reason for the increased hardness (Fasching et al. 2011; Soni et al. 2017).

Gaikwad and his team (2020) found that the cutting rate of the WEDM process was mostly relied on peak current, similarly roughness of the surface was mostly affected by current and voltage. Also, the white layer thickness (WLT) was most influenced by voltage. The WLT was measured from the microscopic images and it is found between 5 and 20 μm. Similarly, researchers (Liu et al. 2016; Manjaiah et al. 2018; Pramanik et al. 2021) made an investigation on the WLT and found that S_{ON} time was the most prominent parameter, WLT was found increased with the increase in the spark on time. But the wire tension was found as the least influenced parameter. The white layer mostly contains the tool elements, oxygen, and carbon.

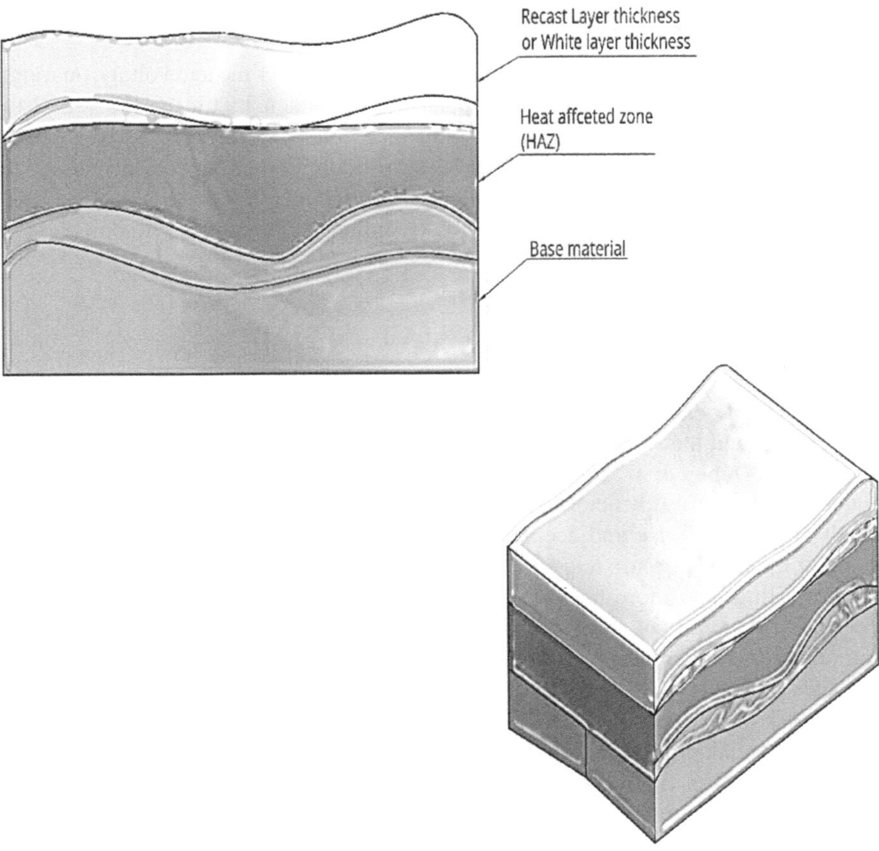

FIGURE 3.3 Pictorial representation of various regions seen in the specimens, machined by using WEDM technique

Based on the literatures two regions were seen in the optical image of the machined surface, i.e., recast/white layer and heat-affected zone. The pictorial representation of the various zones seen in the WEDM machined surface is presented in Figure 3.3.

The impact of the chosen process parameters namely, spark on time, spark off time and voltage was investigated. The ternary element copper and the percentage variation of copper in NiTi alloy showed a notable variation in the surface finish and cutting rate. It was also noted, the voltage primarily influences the percentage of elements in the recast layer. The addition of 20% copper in the NiTi alloy improved the cutting rate. The microhardness was more on the outer surface due to the formation of the recast layer. The maximum thickness of recast layer was measured as 60 μm, the maximum microhardness of 900 Hv was measured in the outer layer, and the micro hardness keeps on decreasing as the depth increases (Manjaiah et al. 2018). Negligible changes on the microhardness were seen after the 60 μm depth. The same trend was noted from most of the work in WEDM (Doreswamy et al. 2022; Fasching et al. 2011; Gaikwad et al. 2020; Liu et al. 2016; Manjaiah et al. 2018; Pramanik et al. 2021; Reddy et al. 2015; Soni et al. 2017).

Similarly, Soni et al. investigated the effects of spark on time and voltage on roughness, cutting speed, microhardness, and surface crack density during the WEDM of Nickel-Titanium-Cobalt (NiTiCo) alloy. Surface cracks were found to be more for higher spark on time and the microhardness was more in the outer surface due to the presence of TiNiCo phases and the same was confirmed by X-ray diffraction (XRD) studies. The presence of newer phases greatly influenced the grain refinement on the surface (Soni et al. 2018).

3.4 LASER BEAM MACHINING (LBM)

Laser beam machining is a futuristic way of machining, performed by the mode of amplified mono chromatic radiation. There is no contact in-between the laser converging lens and the work piece i.e., thermal energy from the laser radiation used for machining (Dubey et al. 2008). This machining process is used for all the form of materials, especially difficult-to-cut materials. This method of machining is commonly used for precise and accurate cutting of materials and drilling micro holes. Some of the hardships are huge initial investment, maintenance cost, safety hazards, poor profiling, and more heat-affected zone in comparison with EDM (Lau et al. 1991). The thickness of recast layer is more compared to the EDM process and formation of taper during the drilling for thicker specimens are some of the other disadvantages (Antar et al. 2016). The schematic representation of the LBM is presented in Figure 3.4.

The essential process of surface expulsion amid the LBM incorporates different stages such as (i) liquifying, (ii) evaporation, and (iii) chemical deterioration (Chemical bonds are broken and the surface material breaks down). Initially, the inert gases/vacuum maintained in the discharge tube, light radiation from the flash lamp excites the electron of the crystal to reach the metastable state. The metastable state is much unstable, in which the electrons last for 10^{-3} seconds after the electron revert to its normal state by losing the energy as a radiation. The laser beam transferred to the specified location by means of the mirror and the same converged to a point by means of a lens. As the laser converged at a point produces huge amount of thermal energy on the surface of the material. The beam produced is very narrow and can be precisely focused with a power density of 1,000 kW/cm^2. The high thermal energy heated the working volume, and converted the surface material into a molten pool, that can be either evaporated, or chemically altered state, so that it can be easily removed by means of a high-pressure assisted gas jet (Antar et al. 2016; Chatterjee et al. 2018; Dubey et al. 2008; Parandoush and Hossain 2014).

LBM has a wide range of applications in the automotive, aircraft, electronics, construction, nuclear, and consumer electronics industries (Dubey et al. 2008). The processing nature of the LBM has made it most suitable for hard and brittle materials. The process parameters affecting the LBM process are laser power, laser wavelength, material thickness, pulse duration, scanning speed, etc., laser power is the most significant parameter during laser drilling process, heat-affected zone increases with the increase in laser power (Chatterjee et al. 2018; Parandoush and Hossain

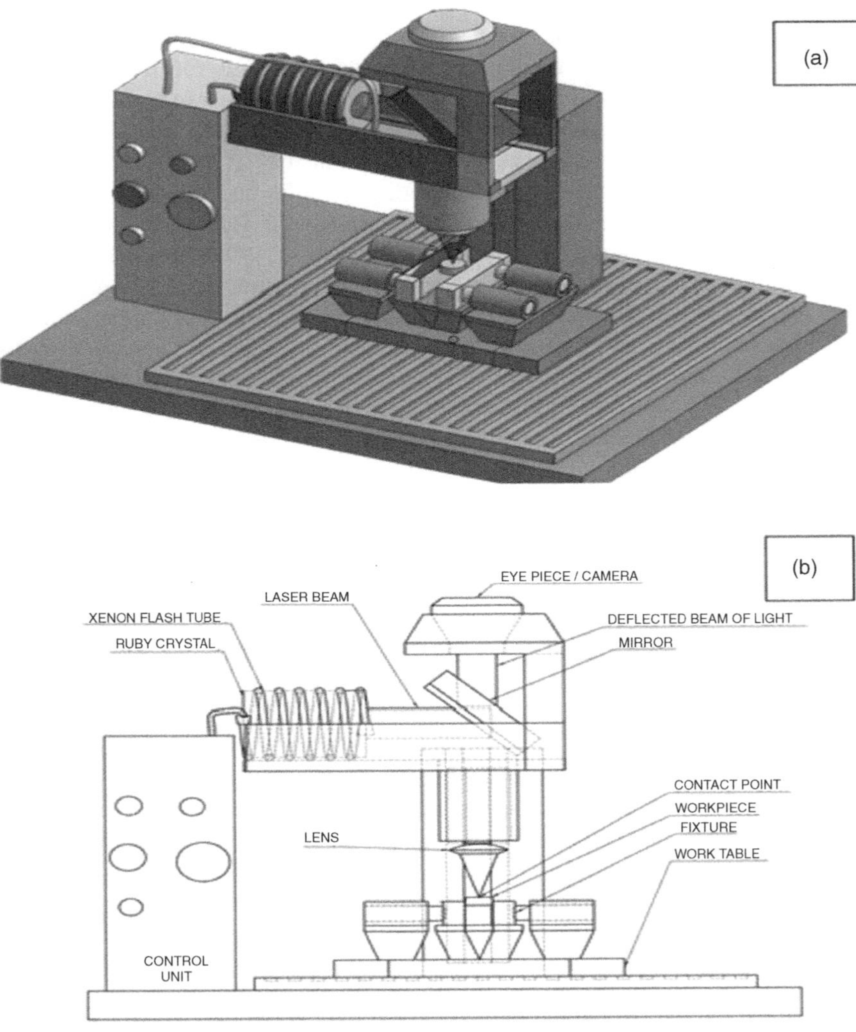

FIGURE 3.4 Basic arrangement of LBM setup, (a) 3D view and (b) schematic view

2014). The drilled hole is narrow tapered with shear ablation zone and the recast zone on the outer surface. The molten region is in the bottom of the hole, further heating turns the material to vaporize (Fan et al. 2016; Li et al. 2021; Uchtmann et al. 2016). The laser holes can be commonly drilled by means of single pulse, percussion, trepanning, and helical trepanning. The trepanning and helical trepanning has been mostly used to drill large-sized holes (i.e., greater than the diameter of beam of laser) (Uchtmann et al. 2016).

Literatures reported that, the effectiveness of the laser machining mostly relied on the laser power, laser wavelength, thickness, feed rate, pulse duration, frequency, type of inert gas, and pressure. LBM has capability to drill hole up to 5 µm i.e., extremely short diameter. The machining characteristics of LBM were reported as hole taper, roughness, recast layer thickness, recast layer phase change, surface defects. This form of thermal based machining technique has been found best suited for high hardness, low thermal conductivity materials like ceramics, nickel, and titanium-built alloys (Dubey et al. 2008; Lau et al. 1991; Uchtmann et al. 2016).

Mathew et al. tried laser machining considering cutting speed, pulse energy, pulse duration, gas pressure as process variable to measure the effect on the output characteristics, heat-affected zone and kerf width (Mathew et al. 1999). Similarly, Ghoreishi et al. was chosen peak power, pulse duration, number of pulses, and assist gas pressure as process parameters. The merit characteristics measured were hole diameter and hole taper (Ghoreishi et al. 2002). Li et al. taken laser current, laser frequency, and cutting speed as machining parameters, width and the heat-affected zone were chosen as merit characteristics (Li et al. 2007). Zadafiya et al. reviewed various non-traditional techniques for machining of nickel and titanium base alloy and reported that the mechanism of disintegration has been found due to the increase in material plasticity because of residual stresses (Zadafiya et al. 2021). Few researchers represented that, commonly in brittle materials, disintegration mechanism is due to the chemical composition change that happened at the interaction area of the laser and the workpiece.

Pradhan et al., made an attempt to study the effect of scanning speed, pulse width, frequency and current on the upper width deviation and lower width deviation during the micromachining of nitinol alloy using Nd-YAG laser. The current has found to be with maximum effect and the pulse frequency has lesser effect. The depth of the deviation increases with pulse width but in contrary, the same decreases with scanning speed. The minimum deviation was seen for the parametric setting of peak current 30 amp, pulse frequency 3,200 Hz, scanning speed 9 mm/s and pulse width 60 ns (Pradhan et al. 2022). Finger et al. investigated the laser machining for faster ablation of the material during the machining of Inconel 718. It is proved that the ablation rate increased with the higher repetition rate. Similarly, the power was proportional to the ablation rate; increase in power more than 57 W increased the ablation rate quickly. The pulse duration has an inverse effect of the ablation rate, as the duration increases above 4 ps, the ablation rate was decreased rapidly. Such a high-speed ablation has a wide application in the fields of aviation and bio medical fields (Finger et al. 2015). The common classification of laser are continuous and pulsed lasers, pulsed lasers are further classified into micros-second laser, nano-second laser, pico-second laser, and femto-second laser (Huang et al. 2004; Leitz et al. 2011). The morphology of the drilled micro hole is presented in Figure 3.5.

The micro machining of the hard materials such as nickel titanium alloys or any super alloys has been found as a difficult task to perform, especially precision and cutting rate are considered. The nanosecond lasers were found efficient, but the problem lied on the deposition of the debris, heat-affected zone, more recast layer development, micro cracks, and delamination of surface material. The usage of femtosecond laser has shown reduced defects and less heat-affected zone compared with nanosecond laser. Das and Pollock, investigated the quality of the micro holes drilled using the femtoseconds laser on nickel superalloy coated with yttria stabilized zirconia. For drilling the expected diameters were 300 µm and 600 µm, so to achieve the

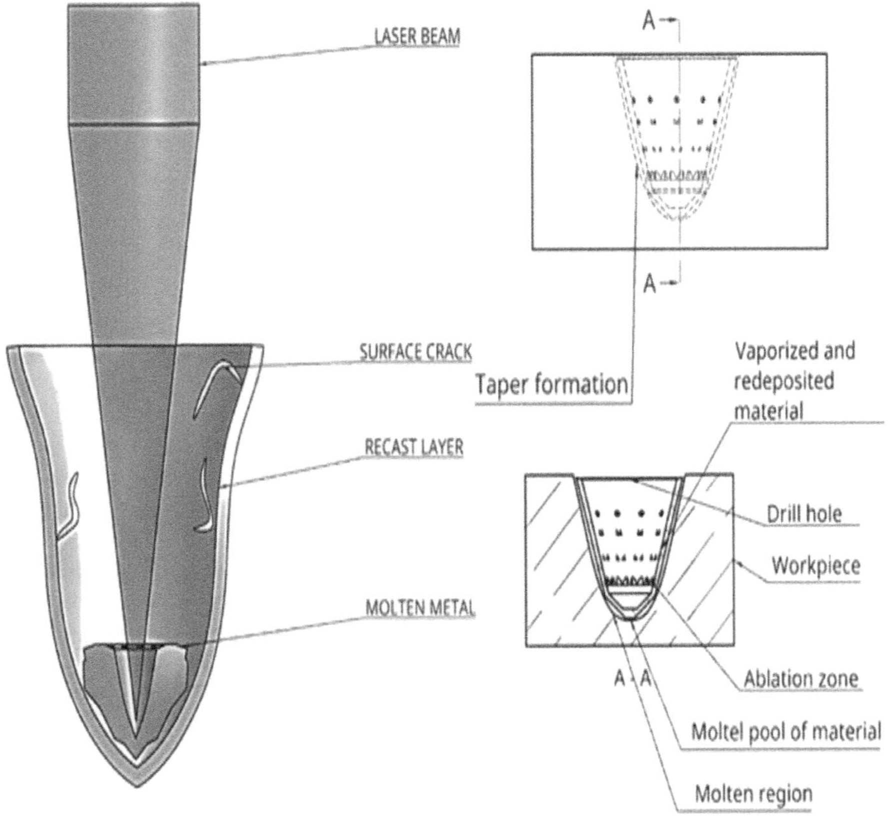

FIGURE 3.5 Pictorial representation of micro hole drilled by femtosecond laser

required diameter trepanning technique was employed. The taper angle of the drilled hole ranging from 3.3° to 5.7°, the ovality of the exit region of hole was more for 1.5 mm thick sample compared with 0.4-mm thick sample. The defects such as recast layer formation, micro crack and delamination of coatings were avoided by using the femtosecond laser (Das and Pollock 2009).

Since the NiTi alloy is a proven biocompatible material, precise machining of such alloy has been done by means of femtosecond laser in order to avoid the surface defects and to retain the phases and biocompatibility. The creation of stents on the NiTi alloy tubes done by means of femtosecond laser, taper during the laser operation majorly governed by pulse duration (Thawari et al. 2005). Huang et al., machined the $Ni_{50.6}Ti_{49.4}$ alloy using the femtosecond laser, they reported that the cutting rate of the machining process directly influenced by the laser power. The cutting depth of the machining increased as the laser power increased. The formation of the recast layer has been seen due to the deposition of the partially melted material, but the thickness is approximately 7 μm, it is very low compared to recast layer formed during the machining done by using the long-pulsed laser. The $Ni_{55.88}Ti_{44.1}$ alloy was machined using the femtosecond laser. The effectiveness of the laser machining relied on the movement route planning (Hung et al. 2015). By which, the characteristics of better-quality surface namely, no recast layer, less heat-affected zone were achieved. The near equiatomic NiTi alloy, the heat-affected zone thickness, material phases, micro hardness were controlled by scanning speed, pulse duration, and pulse width (Pfeifer et al. 2010).

3.5 CONCLUSION

The NiTi alloy is a novel specific purpose material with numerous applications. The properties of the NiTi alloy were found varied by varying the percentage of the Ni and Ti elements in it. The ternary and quaternary element additions are found altering the properties of the NiTi alloy. The machining of NiTi alloy using the conventional technique ends up in surface defects and property change. The non-conventional methods like wire electric discharge machining and laser beam machining were found capable of producing defect free surface and precise machining. The spark ON time, spark OFF time, applied voltage, peak current, wire feed rate, wire tension, dielectric flushing pressure, dielectric medium, and wire material were identified as key parameters for precise machining. Among all, spark ON time, peak current, wire federate, and dielectric pressure were found influencing in surface characteristic of the machined NiTi alloy. The pulsed lasers such as microsecond laser, nanosecond laser, pic-second laser and femtosecond laser were found applied for the micro drilling of the NiTi alloy. Among them, femtosecond laser was found with more accuracy and comparative dimensional accuracies even in helical trepan drilling. The same can be controlled by varying scanning speed, pulse energy, etc. Considering the features and benefits of these two considered modern machining processes, it can be said that modern machining processes have a potential for sustainable machining of difficult-to-machine materials.

REFERENCES

Abdulkareem, Suleiman, Ahsan Ali Khan, and Mohamed Konneh. 2010. "Cooling effect on electrode and process parameters in EDM." *Materials and Manufacturing Processes* 25 (6). Taylor & Francis: 462–466.

Andani, Mohsen Taheri, Soheil Saedi, Ali Sadi Turabi, M. R. Karamooz, Christoph Haberland, Haluk Ersin Karaca, and Mohammad Elahinia. 2017. "Mechanical and shape memory properties of porous Ni50. 1Ti49. 9 alloys manufactured by selective laser melting." *Journal of the Mechanical Behavior of Biomedical Materials* 68. Elsevier: 224–231.

Antar, Mohammad, Dimitrios Chantzis, Sundar Marimuthu, and Philip Hayward. 2016. "High speed EDM and laser drilling of aerospace alloys." *Procedia CIRP* 42. Elsevier: 526–531.

Arunachalam, R., and M. A. Mannan. 2000. "Machinability of nickel-based high temperature alloys." *Machining Science and Technology* 4 (1). Taylor & Francis: 127–168.

Bahraminasab, Marjan, and Ali Jahan. 2011. "Material selection for femoral component of total knee replacement using comprehensive VIKOR." *Materials & Design* 32 (8–9). Elsevier: 4471–4477.

Barras, Christen. D. J., and Kari Anne Myers. 2000. "Nitinol–its use in vascular surgery and other applications." *European Journal of Vascular and Endovascular Surgery* 19 (6). Elsevier: 564–569.

Behera, Ajit, Dipen Kumar Rajak, Reza Kolahchi, Maria-Luminiţa Scutaru, and Catalin I. Pruncu. 2020. "Current global scenario of Sputter deposited NiTi smart systems." *Journal of Materials Research and Technology* 9 (6). Elsevier: 14582–14598.

Chatterjee, Suman, Siba Sankar Mahapatra, Arpan Mondal, and Kumar Abhishek. 2018. "An experimental study on drilling of titanium alloy using CO_2 laser." *Sādhanā* 43. Springer: 1–14.

Chaudhari, Rakesh, Jay J. Vora, Vivek Patel, L. N. López de Lacalle, and D. M. Parikh. 2020. "Effect of WEDM process parameters on surface morphology of nitinol shape memory alloy." *Materials* 13 (21). MDPI: 4943.

Choudhary, Nitin, and Davinder Kaur. 2016. "Shape memory alloy thin films and hetero-structures for MEMS applications: A review." *Sensors and Actuators A: Physical* 242. Elsevier: 162–181.

Daneshmand, Saeed, Ehsan Farahmand Kahrizi, Esmail Abedi, and M. Mir Abdolhosseini. 2013. "Influence of machining parameters on electro discharge machining of NiTi shape memory alloys." *International Journal of Electrochemical Science* 8 (3). ESG publisher: 3095–3104.

Das, Dipak K., and Tresa M. Pollock. 2009. "Femtosecond laser machining of cooling holes in thermal barrier coated CMSX4 superalloy." *Journal of Materials Processing Technology* 209 (15–16). Elsevier: 5661–5668.

Dash, Barsharani, Manojit Das, Monalisa Das, Trupti R. Mahapatra, and Debadutta Mishra. 2019. "A concise review on machinability of NiTi shape memory alloys." *Materials Today: Proceedings* 18. Elsevier: 5141–5150.

Doreswamy, Deepak, D. Sai Shreyas, Subraya Krishna Bhat, and Rajath N. Rao. 2022. "Optimization of material removal rate and surface characterization of wire electric discharge machined Ti-6Al-4V alloy by response surface method." *Manufacturing Review* 9. EDP Sciences: 15.

Dubey, Avanish Kumar, and Vinod Yadava. 2008. "Laser beam machining—A review." *International Journal of Machine Tools and Manufacture* 48 (6). Elsevier: 609–628.

Elloy, M. A., J. T. M. Wright, and M. E. Cavendish. 1976. "The basic requirements and design criteria for total joint prostheses." *Acta Orthopaedica Scandinavica* 47 (2). Taylor & Francis: 193–202.

Ezugwu, E. O., R. B. Da Silva, J. Bonney, and A. R. Machado. 2005. "Evaluation of the performance of CBN tools when turning Ti–6Al–4V alloy with high pressure coolant supplies." *International Journal of Machine Tools and Manufacture* 45 (9). Elsevier: 1009–1014.

Fan, Zhengjie, Xia Dong, Kedian Wang, Wenqiang Duan, Rujia Wang, Xuesong Mei, Wenjun Wang, Jianlei Cui, Xin Yuan, and Chengying Xu. 2016. "Effect of drilling allowance on TBC delamination, spatter and re-melted cracks characteristics in laser drilling of TBC coated superalloys." *International Journal of Machine Tools and Manufacture* 106. Elsevier: 1–10.

Fasching, Audrey, D. Norwich, T. Geiser, and Graeme W. Paul. 2011. "An evaluation of a NiTiCo alloy and its suitability for medical device applications." *Journal of Materials Engineering and Performance* 20 (4–5). Springer: 641–645.

Finger, J., C. Kalupka, and M. Reininghaus. 2015. "High power ultra-short pulse laser ablation of IN718 using high repetition rates." *Journal of Materials Processing Technology* 226. Elsevier: 221–227.

Gaikwad, Mahendra Uttam, A. Krishnamoorthy, and Vijaykumar S. Jatti. 2020. "Investigation on effect of process parameter on surface integrity during electrical discharge machining of NiTi 60." *Multidiscipline Modeling in Materials and Structures* 16 (6). Emerald Publishing Limited: 1385–1394.

Ghoreishi, M., D. K. Y. Low, and L. Li. 2002. "Comparative statistical analysis of hole taper and circularity in laser percussion drilling." *International Journal of Machine Tools and Manufacture* 42 (9). Elsevier: 985–995.

Goyal, Ashish, and H. U. Z. E. F. Ur Rahman. 2021. "Experimental studies on Wire EDM for surface roughness and kerf width for shape memory alloy." *Sādhanā* 46 (3). Springer: 160.

Gupta, Deepak Kumar, and Avanish Kumar Dubey. 2021. "Multi process parameters optimization of Wire-EDM on shape memory alloy (Ni54. 1Ti) using Taguchi approach." *Materials Today: Proceedings* 44. Elsevier: 1423–1427.

Hassan, M. R., Mershad Mehrpouya, and S. Dawood. 2014. "Review of the machining difficulties of nickel-titanium based shape memory alloys." *Applied Mechanics and Materials* 564. Trans Tech Publications Ltd.: 533–537.

Hodgson, Norman, Albrecht Steinkopff, Sebastian Heming, Hortense Allegre, Hatim Haloui, Tony S. Lee, Michael Laha, and Joris VanNunen. 2021. "Ultrafast laser machining: Process optimization and applications." *Laser Applications in Microelectronic and Optoelectronic Manufacturing (LAMOM) XXVI* 11673. SPIE: 21–41.

Hornbuckle, B. Chad, X. Yu Xiao, Ronald D. Noebe, Richard Martens, Mark L. Weaver, and Gregory B. Thompson. 2015. "Hardening behavior and phase decomposition in very Ni-rich Nitinol alloys." *Materials Science and Engineering: A* 639. Elsevier: 336–344.

Huang, H., H. Y. Zheng, and G. C. Lim. 2004. "Femtosecond laser machining characteristics of Nitinol." *Applied Surface Science* 228 (1–4). Elsevier: 201–206.

Hung, Chia-Hung, Fuh-Yu Chang, Tien-Li Chang, Yu-Ting Chang, Kai-Wen Huang, and Po-Chin Liang. 2015. "Micromachining NiTi tubes for use in medical devices by using a femtosecond laser." *Optics and Lasers in Engineering* 66. Elsevier: 34–40.

Jilani, S. Tariq, and P. C. Pandey. 1984. "Experimetnal investigations into the performance of water as dielectric in EDM." *International Journal of Machine Tool Design and Research* 24 (1) Elsevier: 31–43.

Kanlayasiri, K., and S. Boonmung. 2007. "Effects of wire-EDM machining variables on surface roughness of newly developed DC 53 die steel: Design of experiments and regression model." *Journal of Materials Processing Technology* 192. Elsevier: 459–464.

Khanlari, Khashayar, Maziar Ramezani, and Piaras Kelly. 2018. "60NiTi: A review of recent research findings, potential for structural and mechanical applications, and areas of continued investigations." *Transactions of the Indian Institute of Metals* 71. Springer: 781–799.

Kulkarni, Vinayak N., V. N. Gaitonde, S. R. Karnik, M. Manjaiah, and J. Paulo Davim. 2020. "Machinability analysis and optimization in wire EDM of medical grade NiTiNOL memory alloy." *Materials* 13 (9). MDPI: 2184.

Kulkarni, Vinayak N., V. N. Gaitonde, Viranna Hadimani, and Vasant Aiholi. 2018. "Analysis of wire EDM process parameters in machining of NiTi superelastic alloy." *Materials Today: Proceedings* 5 (9). Elsevier: 19303–19312.

Kumar, Siva, M. Adam Khan, and B. Muralidharan. 2019. "Processing of titanium-based human implant material using wire EDM." *Materials and Manufacturing Processes* 34 (6). Taylor & Francis: 695–700.

Lau, W. S., and Wing Bun Lee. 1991. "A comparison between edivi wire-cut and laser cutting of carbon fibre composite materials." *Material and Manufacturing Process* 6 (2). Taylor & Francis: 331–342.

Leitz, Karl-Heinz, Benjamin Redlingshöfer, Yvonne Reg, Andreas Otto, and Michael Schmidt. 2011. "Metal ablation with short and ultrashort laser pulses." *Physics Procedia* 12. Elsevier: 230–238.

Li, Chen-Hao, Ming-Jong Tsai, and Ciann-Dong Yang. 2007. "Study of optimal laser parameters for cutting QFN packages by Taguchi's matrix method." *Optics & Laser Technology* 39 (4). Elsevier: 786–795.

Li, Qiang, Yanjun Zeng, and Xiaoying Tang. 2010. "The applications and research progresses of nickel–titanium shape memory alloy in reconstructive surgery". *Australasian Physical & Engineering Sciences in Medicine* 33. Springer: 129–136.

Li, Wenyuan, Yu Huang, Xinghua Chen, Guojun Zhang, Youmin Rong, and Ya Lu. 2021. "Study on laser drilling induced defects of CFRP plates with different scanning modes based on multi-pass strategy." *Optics & Laser Technology* 144. Elsevier: 107400.

Liu, J. F., Y. B. Guo, T. M. Butler, and M. L. Weaver. 2016. "Crystallography, compositions, and properties of white layer by wire electrical discharge machining of nitinol shape memory alloy." *Materials & Design* 109. Elsevier: 1–9.

Maher, Ibrahem, Ahmed A. D. Sarhan, and M. Hamdi. 2015. "Review of improvements in wire electrode properties for longer working time and utilization in wire EDM machining." *The International Journal of Advanced Manufacturing Technology* 76. Springer: 329–351.

Mandal, Amitava, and Amit Rai Dixit. 2014. "State of art in wire electrical discharge machining process and performance." *International Journal of Machining and Machinability of Materials* 16 (1). Inderscience: 1–21.

Manjaiah, M., S. Narendranath, and S. Basavarajappa. 2014. "Review on non-conventional machining of shape memory alloys." Transactions of Nonferrous Metals Society of China 24 (1). Elsevier: 12–21.

Manjaiah, M., S. Narendranath, S. Basavarajappa, and V. N. Gaitonde. 2018. "Investigation on material removal rate, surface and subsurface characteristics in wire electro discharge machining of Ti50Ni50-xCux shape memory alloy." *Proceedings of the Institution of Mechanical Engineers, Part L: Journal of Materials: Design and Applications* 232 (2). SAGE Publishing: 164–177.

Matassi, Fabrizio, Alessandra Botti, Luigi Sirleo, Christian Carulli, and Massimo Innocenti. 2013. "Porous metal for orthopedics implants." *Clinical Cases in Mineral and Bone Metabolism* 10 (2). National Library of Medicine, PubMed Central: 111–115.

Mathew, Jose, G. L. Goswami, N. Ramakrishnan, and N. K. Naik. 1999. "Parametric studies on pulsed Nd: YAG laser cutting of carbon fibre reinforced plastic composites." *Journal of Materials Processing Technology* 89. Elsevier: 198–203.

Minshull, James P., Steffen Neumeier, Mattew G. Tucker, and Howard James Stone. 2011. "A1-L12 structures in the Al-Co-Ni-Ti quaternary phase system." *Advanced Materials Research* 278. Trans Tech Publications Ltd: 399–404.

Mohammed, M. A., A. A. Aljubouri, and S. H. Mohammed. 2020. "Preparation and characterization of equiatomic NiTi shape memory alloy." *In IOP Conference Series: Materials Science and Engineering* 757 (1). IOP Publishing: 012059.

Naresh, C., P. S. C. Bose, and C. S. P. Rao. 2016. "Shape memory alloys: A state of art review." *IOP Conference Series: Materials Science and Engineering* 149 (1). IOP Publishing: 012054.

Nishanth, B. S., Vinayak N. Kulkarni, and V. N. Gaitonde. 2019. "A review on conventional and non-conventional machining of titanium and nickel based alloys." *AIP Conference Proceedings* 2200 (1). AIP Publishing LLC: 020091.

Parandoush, Pedram, and Altab Hossain. 2014. "A review of modeling and simulation of laser beam machining." *International Journal of Machine Tools and Manufacture* 85. Elsevier: 135–145.

Patel, Swadhin Kumar, Bikram Behera, Biswajit Swain, Rakesh Roshan, Deepak Sahoo, and A. Behera. 2020. "A review on NiTi alloys for biomedical applications and their biocompatibility." *Materials Today: Proceedings* 33. Elsevier: 5548–5551.

Paula, A. S., K. K. Mahesh, C. M. L. Dos Santos, F. M. Braz Fernandes, and C. S. da Costa Viana. 2008. "Thermomechanical behavior of Ti-rich NiTi shape memory alloys." *Materials Science and Engineering: A* 481. Elsevier: 146–150.

Pfeifer, Ronny, Dirk Herzog, Michael Hustedt, and Stephan Barcikowski. 2010. "Pulsed Nd: YAG laser cutting of NiTi shape memory alloys—Influence of process parameters." *Journal of Materials Processing Technology* 210 (14). Elsevier: 1918–1925.

Pradhan, Subhadip, Piyush Bhusan Dash, Kanchan Kumari, and Debabrata Dhupal. 2022. "Study of micro machining characteristics by Nd-YAG Laser on NiTinol shape memory alloy." *Advances in Materials and Processing Technologies* 8 (2). Taylor & Francis: 670–684.

Pramanik, A., A. K. Basak, C. Prakash, S. Shankar, Shubham Sharma, and S. Narendranath. 2021. "Recast layer formation during wire electrical discharge machining of Titanium (Ti-Al6-V4) alloy." *Journal of Materials Engineering and Performance* 30 (12). Springer: 8926–8935.

Qadir, Razaw, Safar Mohammed, K. Ö. K. Mediha, and Ibrahim Qader. 2021. "A review on NiTiCu shape memory alloys: Manufacturing and characterizations." *Journal of Physical Chemistry and Functional Materials* 4 (2). DergiPark: 49–56.

Reddy, V. Vikram, P. Madar Valli, A. Kumar, and Ch Sridhar Reddy. 2015. "Influence of process parameters on characteristics of electrical discharge machining of PH17-4 stainless steel." *Journal of Advanced Manufacturing Systems* 14 (3). World Scientific Publishing Co: 189–202.

Sharma, Neeraj, Tilak Raj, and Kamal Jangra. 2015. "Applications of nickel-titanium alloy." *Journal of Engineering and Technology* 5 (1). Online Jet: 1.

Singh, Ranjit, Ravi Pratap Singh, and Rajeev Trehan. 2021a. "State of the art in processing of shape memory alloys with electrical discharge machining: A review." *Proceedings of the Institution of Mechanical Engineers, Part B: Journal of Engineering Manufacture* 235 (3). SAGE Publishing: 333–366.

Singh, Shalini, N. Resnina, S. Belyaev, A. N. Jinoop, Ashish Shukla, I. A. Palani, C. P. Paul, and K. S. Bindra. 2021b. "Investigations on NiTi shape memory alloy thin wall structures through laser marking assisted wire arc based additive manufacturing." *Journal of Manufacturing Processes* 66. Elsevier: 70–80.

Slătineanu, Laurenţiu, Oana Dodun, Margareta Coteaţă, Gheorghe Nagîţ, Irina Beşliu Băncescu, and Adelina Hriţuc. 2020. "Wire electrical discharge machining—A review." *Machines* 8 (4). MDPI: 69.

Soni, Hargovind, Narendranath Sannayellappa, and Ramesh Motagondanahalli Rangarasaiah. 2017. "An experimental study of influence of wire electro discharge machining parameters on surface integrity of TiNiCo shape memory alloy." *Journal of Materials Research* 32 (16). Cambridge University Press: 3100–3108.

Soni, Hargovind, S. Narendranath, and M. R. Ramesh. 2018. "Evaluation of wire electro discharge machining characteristics of Ti50Ni45Co5 shape memory alloy." *Pulse* 1 (2). COPEN Comprehensive Portal: 3.

Tadayyon, Ghazal, Mohammad Mazinani, Yina Guo, Seyed Mojtaba Zebarjad, Syed A. M. Tofail, and Manus J. P. Biggs. 2016. "Study of the microstructure evolution of heat treated Ti-rich NiTi shape memory alloy." *Materials Characterization* 112. Elsevier: 11–19.

Tadayyon, Ghazal, Yina Guo, Mohammad Mazinani, Seyed Mojtaba Zebarjad, Peter Tiernan, Syed A. M. Tofail, and Manus J. P. Biggs. 2017. "Effect of different stages of deformation on the microstructure evolution of Ti-rich NiTi shape memory alloy." *Materials Characterization* 125. Elsevier: 51–66.

Tanzi, M. C., S. Farè, and G. Candiani. 2019. "Foundations of biomaterials engineering". *Academic Press*, Elsevier.

Thawari, G., J. K. Sarin Sundar, G. Sundararajan, and S. V. Joshi. 2005. "Influence of process parameters during pulsed Nd: YAG laser cutting of nickel-base superalloys." *Journal of Materials Processing Technology* 170 (1–2). Elsevier: 229–239.

Uchtmann, Hermann, Chao He, and Arnold Gillner. 2016. "High precision and high aspect ratio laser drilling: Challenges and solutions." *In High-Power Laser Materials Processing: Lasers, Beam Delivery, Diagnostics, and Applications* V 9741. SPIE: 36–47.

Uthayakumar, M., M. Adam Khan, S. Thirumalai Kumaran, Adam Slota, and Jerzy Zajac. 2016. "Machinability of nickel-based superalloy by abrasive water jet machining." *Materials and Manufacturing Processes* 31 (13). Taylor & Francis: 1733–1739.

Valaki, Janak B., and Pravin P. Rathod. 2016. "Assessment of operational feasibility of waste vegetable oil based bio-dielectric fluid for sustainable electric discharge machining (EDM)." *The International Journal of Advanced Manufacturing Technology* 87. Springer: 1509–1518.

Wang, Xiebin, Sergey Kustov, and Jan Van Humbeeck. 2018. "A short review on the microstructure, transformation behavior and functional properties of NiTi shape memory alloys fabricated by selective laser melting." *Materials* 11 (9). MDPI: 1683.

Zadafiya, Kishan, Soni Kumari, Suman Chatterjee, and Kumar Abhishek. 2021. "Recent trends in non-traditional machining of shape memory alloys (SMAs): A review." *CIRP Journal of Manufacturing Science and Technology* 32. Elsevier: 217–227.

Zhang, Xuping, Guiji Wang, Binqiang Luo, Simon N. Bland, Fuli Tan, Feng Zhao, Jianheng Zhao, Chengwei Sun, and Cangli Liu. 2018. "Mechanical response of near-equiatomic NiTi alloy at dynamic high pressure and strain rate." *Journal of Alloys and Compounds* 731. Elsevier: 569–576.

Zhu, Jianing, Qunfeng Zeng, and Tao Fu. 2019. "An updated review on TiNi alloy for biomedical applications." *Corrosion Reviews* 37 (6). Walter de Gruyter: 539–552.

4 Eco-Friendly Dielectrics for EDM and WEDM Processes

Mustafa Günay and Ahmet Tolunay Işık

4.1 INTRODUCTION

Problems such as tool damage, tolerance, surface quality, and machining difficulties have emerged in the use of conventional machining methods with the developing material technology (Deshmukh et al., 2022). In the industrial sense, innovative technologies also called advanced manufacturing methods or unconventional machining methods are used to solve these problems. Unconventional machining methods have various technologies that remove material from the workpiece using various forms of energy (Groover, 2010). For example, manufacturing technologies such as electrical discharge machining (EDM), abrasive water jet machining (AWJM), electron beam machining (EBM), laser machining (LM), electrochemical machining (ECM), and ultrasonic machining (UM) are widely used (Rohith et al., 2019). Among these methods, EDM is an indispensable technology in the production of complex or simple geometry, macro, or nano-scale parts from materials (Electrically conductive) that are difficult to process with conventional methods (Gupta et al., 2015; Jahan, 2019). EDM is a method in which electrical sparks created between two conductors with the help of a direct current pulse generator remove the material thermo-electrically. In the EDM process, the dielectric fluid takes on the task of cooling and cleaning the processing zone. The electrostatic field consisting of high current and voltage causes disruption in the dielectric medium, causing the movement of sparks (ions and electrons) between the electrodes. This event happens in a few microseconds (Çakıroğlu, 2022). However, as with any manufacturing method, electrical discharge machining has some disadvantages and advantages. The advantages of this manufacturing method can be listed as the fact that hardness and brittleness are not criteria for machining, complex shapes can be processed with high precision, non-contact machining without cutting forces, no burr formation, a good surface quality that does not require polishing, low tooling cost, total cost advantage (special tooling, finishing process, etc.). The disadvantages are high energy use, conductivity requirement, low machining speed, tool wear, excessive cutting, crack, and white layer formation, complex machining process (shaping the tool, dielectric fluid and acting, parameter selection, etc.) and the negative impact of dielectrics (Qudeiri et al., 2020). Among these disadvantages, dielectrics, which negatively affect the health of the operator and the environment, are frequently used in EDM operations to increase

DOI: 10.1201/9781003352402-4

the processing speed and to minimize tool wear (Niamat et al., 2017). During electrical discharge machining, a fume is released from the dielectric during material removal by thermal energy. Harmful aerosols and gases in this smoke can cause serious health problems due to inhalation and skin contact (Leppert, 2018; Zia et al., 2019). In addition, dielectrics that are not naturally degradable cause harm to the environment as waste. In order to eliminate the negative effects of dielectrics, some researches have been done in the past. This chapter mainly covers the application and sustainability of clean and environmentally friendly dielectrics in EDM. Thus, it is aimed that the mentioned machining method is safe and environmentally friendly for future generations.

4.2 ELECTRICAL DISCHARGE MACHINING (EDM)

Electrical discharge machining (EDM) is one of the widely used unconventional machining processes. EDM is an electro-thermal machining process in which material removal is performed with the thermal energy of the spark by creating a spark in a high-frequency discharge current (Akıncıoğlu, 2022; Khilji et al., 2022). Figure 4.1 presents the mechanism of material removal in EDM. Two conductive electrodes for sparking; a potential difference is applied between the electrode/tool (cathode) and the workpiece (anode). The intense thermal flux that occurs locally can cause an instantaneous temperature increase of up to 12,000°C. Such an increase in temperature allows the material to be removed by melting or evaporation. Processing takes place in a continuously circulating dielectric to remove residual materials called debris (Equbal and Sood, 2014).

Since EDM is a non-contact machining method, overcut zones occur during processing. Gap and overcut zones are the areas that allow sparks to occur between the tool and the workpiece during machining. The overcut zone is the name given to the area between the tool corner and the workpiece wall. The gap distance is the gap between the tool machining surface and the workpiece surface (Figure 4.2). Short circuits and irregular spark pulses can occur during machining if the gap is too short. During processing, the material is first removed from the areas closest to the spark distance. This process continues repeatedly until the machining time or distance is exhausted. As the gap distance increases, the area to generate the spark decreases and a rougher operation occurs on the surface (Jameson, 2001). The lower conductivity limit for EDM is 2×102 $\Omega^{-1}m^{-1}$, above which it is impossible to create a spark. Therefore, EDM of ceramics and their composites may be possible if they have a conductive matrix to conduct the discharge current (Razzell, 2000).

After the EDM process, it is generally seen that three different layers are formed (Figure 4.3). The electro-erosion layer on the top, which can be easily removed, consists of the tool electrode adhering to the workpiece surface and the processed material. The next layer, the white layer, consists of resolidified and non-ejectable molten electrode (tool and workpiece) materials containing dielectric material (carbon). In the pulse interval, the molten material solidifies by being rapidly cooled by the dielectric. This causes the white layer to harden and become brittle, and this cycle can create microcracks in the structure. White layer thickness can reduce workpiece life but improve corrosion and wear resistance. It was determined that the carbon in

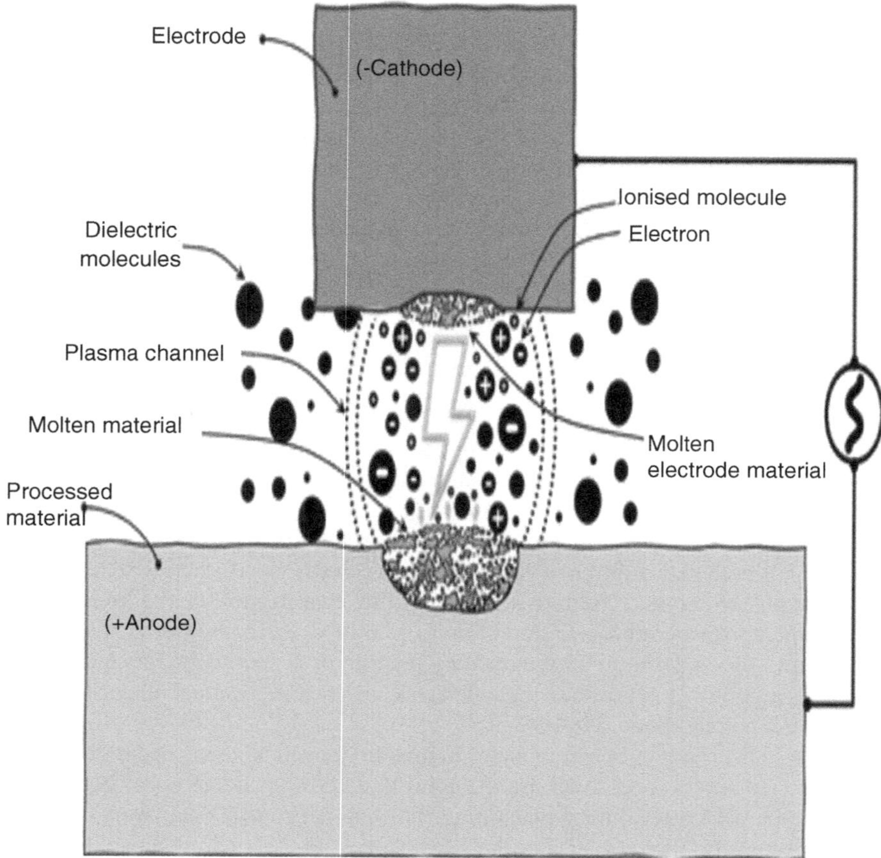

FIGURE 4.1 Processing mechanisms in conventional EDM

the white layer separated from the carbon-based dielectric and penetrated the work-piece (by pyrolysis), while oxygen penetrated the workpiece in water-based dielectrics. The final layer, the heat affected zone (HAZ), is defined as the annealed zone without melting. The depth of the white layer and HAZ layers varies according to the material's specific heat value and impact energy (Kumar et al., 2009; Prakash et al., 2018).

4.3 EDM METHODS

Electrical discharge machining technique can be applied for many different processes such as sinking, cutting and grinding. In addition to these methods, hybrid EDM applications such as dry, powder mixed, rotating electrode, and ultrasonic vibration can be used (Mohd et al., 2007). However, the most widely used in various applications in many fields of industry are die-sinking and wire EDM methods.

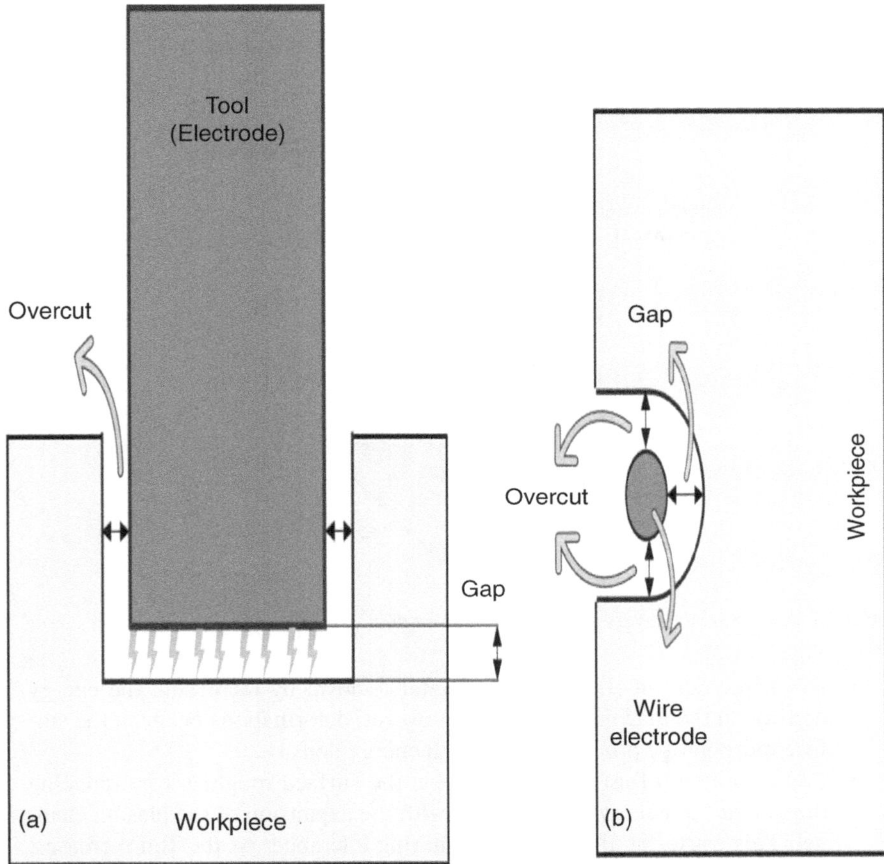

FIGURE 4.2   Overcut and gap zones, (a) Die-sinking EDM and (b) Wire EDM

4.3.1   DIE-SINKING EDM

The classical EDM also called die-sinking or plunge EDM is the process of transferring the tool shape to the workpiece by means of rhythmic sparks. In the EDM machine, there should be power supply system, the chamber where the machining takes place (workpiece and tool), tool feeding system (servo mechanism), and dielectric feeding unit as the main elements. Various conductive materials can be used as tool (electrode) material. For example, copper and graphite materials are widely used as electrodes. In some cases, it may be necessary to use a new electrode for a very good surface finish due to tool wear on the erosion bench. On the other hand, this change may not be necessary due to the positive effect of hydrocarbon dielectrics on surface roughness and tool wear (Bisaria and Shandilya, 2015). Machining quality is related to electrical and non-electrical parameters used in plunge electro-erosion machining. Some of these parameters and their effects are given below.

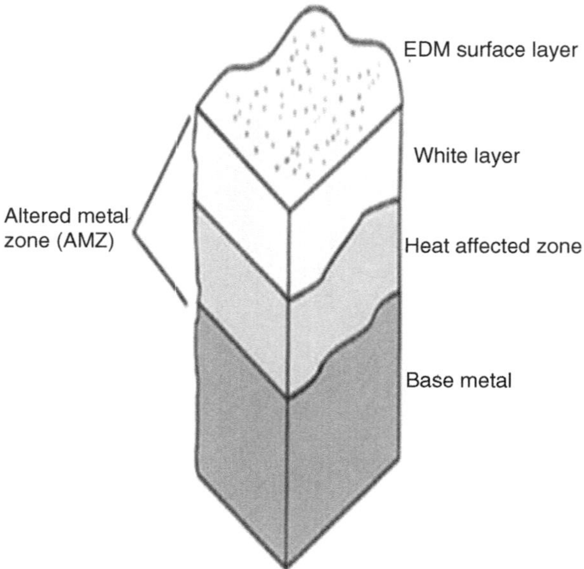

FIGURE 4.3   Surface layers after EDM (Kumar et al., 2009)

- Discharge current (I) accelerates metal removal by increasing the energy density on the machining surface. However, deformations occur in the surface morphology processed with high energy density.
- Pulse duration (Ton) positively affects the surface roughness by reducing the amount of energy per unit area with the expansion of the plasma channel. This causes an increase in machining tolerance. As the Ton increases, carbon adhering from the dielectric to the electrode during machining increases material hardness or prevents tool wear.
- Pulse interval (Toff) provides the discharge of the debris and the realization of de-ionization with the help of dielectric by interrupting the discharges.
- Reference voltage (V) is specified as the range of motion of the spark in the processing zone. High voltage range reduces the processing speed, while the low voltage causes short circuits, negatively affecting the processing quality.
- Open circuit voltage (Voc) can also be expressed as ignition delay. It affects the size of the rash.
- Polarity is characterized as the charge difference between the electrodes. Generally, the mobility of electrons (current) is from the negatively charged (−) electrode to the positively charged (+) electrode.
- Washing pressure (P) is required to move debris from the processing zone and control the plasma channel. At very high pressures, the electric field narrows and the gas bubbles that carry the debris become smaller. While this situation causes low machining efficiency, it brings with it the problems of cleaning debris at low pressures and short-circuiting in the pulse range.

- Dielectric type determines the properties required for debris removal and cooling of the machining cavity. The material to be machined should be selected by the tool material and machining conditions (Almeida et al., 2021; Koyano et al., 2016; Ramasawmy and Blunt, 2004).

4.3.2 Wire EDM

In the wire WEDM, the conductive wire with a small diameter independent of the workpiece shape in the dielectric moves continuously on a defined path for the purpose of cutting the workpiece. Wire electrodes are made of conductive metals, usually 0.05–0.30 mm in size (Ho et al., 2004, Reddy et al., 2021). In Figure 4.4, the machining process in WEDM is shown schematically.

Sudden thermal changes (heating and cooling) during WEDM affect the hardness of the machined surface at a certain depth, making the workpiece surface brittle. This causes the formation of microcracks on the EDM surface. In order to minimize these effects, process parameters, wire properties and dielectric type must be set correctly (Asgar and Singholi, 2018).

In wire EDM, the machining quality is mostly related to the machining parameters. As with the plunge EDM, the effects of I, Ton, Toff, V, P and dielectric type are similar. In this method, the machining parameters and their effects, which are highly important in terms of part quality, are summarized below.

- When the wire speed is high, metal removal rate (MRR) decreases with the decrease in the heat in the machining area, and breaks due to stresses may occur. It is necessary to adjust the wire speed at the optimum level, since breaks due to wear are observed at low wire speed.
- While wire tension may cause breakage, machining errors can occur at low tension, short circuits or wire oscillations at high pressure. In this respect, it is expected that the wire tension should be adjusted at the optimum level.
- Wire or tool type is seen as a parameter that directly affects the quality of the machining. It should be selected considering the material type and performance criteria as well as the dielectric type. Electrodes that will react with the dielectric or the workpiece material should not be selected in terms of machining accuracy, environmental and operator health (Garg et al., 2010; Kumar et al., 2011).

Accordingly, MRR, tool/wire wear rate (T/WWR), surface quality (SQ) are generally considered as performance criteria in machining with both EDM methods, and these are particularly affected by the changes in electrode material, dielectric, and electrical parameters (Gupta and Kumar, 2021).

4.4 IMPACTS OF DIELECTRICS ON EDM PERFORMANCE

Removal of debris formed during EDM and insulation between electrodes are required. The potential difference created between the electrodes ionizes the dielectric, allowing the movement of sparks in the processing area for EDM (Pei et al.,

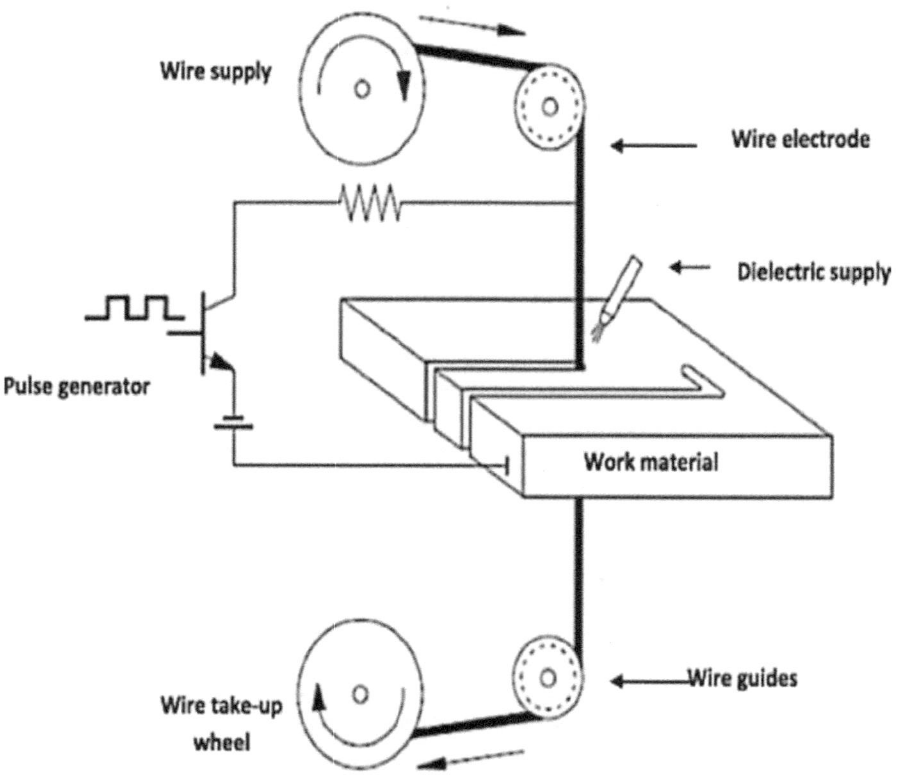

FIGURE 4.4   Cutting process in WEDM (Goyal, 2017)

2017). Therefore, EDM systems use dielectric as the working medium. The main tasks of dielectrics are given below:

- Removal of the spillage from the cavity,
- Cooling of worn and overheated machining points,
- Ensuring ionization in the space,
- Insulation of tool and workpiece electrodes,
- Narrowing of the discharge channel in order to increase the energy density.

Dielectric fluids specially adapted to EDM requirements can be used. In terms of cost, hydrocarbon compounds in the form of mineral oil or synthetic products and pure water are preferred as dielectrics. Deionized water is generally used in wire-EDM processes due to the low effect of tool wear on the machining area and micro-EDM processes due to the small molecular structure in volume (Kumar and Dhinesh, 2020; Marashi et al., 2016).

During EDM, some undesirable gases emerge from dielectric liquids that pollute the environment and affect operators' health (Gupta and Gupta, 2019). Considering that the dielectric is the main parameter for the surface roughness and integrity of the liquid, it has been seen that the use of water-based or gaseous dielectrics is more suitable for the environment and health. Alternatively, using gas dielectrics can be preferred to minimize the abovementioned adverse effects (Dhakar et al., 2019; Leão and Pashby, 2004; Shirguppikar and Dabade, 2018). The dielectric must be selected according to the workpiece properties, machining tolerance, machining speed, desired surface quality, and the type of tool to be used. The ideal dielectric properties for EDM can be listed as follows (Valaki et al., 2016).

- Density can change with the molecules mixed into the dielectric during processing and the evaporation of the volatiles in the dielectric. Low-density dielectrics can carry more debris under the influence of hydrodynamic forces.
- Low viscosity is required in narrow machining space to remove debris and cooling in finishing. However, high-viscosity dielectrics are recommended for roughing or high-speed machining. Also, a more viscous liquid absorbs a significant portion of the smoke and vapor with its increased dielectric level.
- High flash or fire point is necessary to prevent undesirable fires under operating, transportation, and storage conditions.
- Dielectrics must have the necessary dielectric strength to provide insulation between the electrodes up to the breakdown voltage. At low dielectric strength, the spark distance between the electrodes is reduced, resulting in debris and short circuits that cannot be cleaned.
- Gases and aerosols that will adversely affect the environment and operator health are released during processing in the toxic and corrosive content.
- Being chemically inert is important for the longevity and protection of the dielectric to be used.
- High thermal conductivity allows rapid heat transfer in the processing zone, improving processing speed.
- High specific heat absorbs heat energy in sudden temperature changes and ensures effective cooling in the spark gap.

In industrial applications, results are difficult to predict and control due to the complex influence of many parameters in EDM. This often results in ignoring some of the disadvantages for EDM, such as the use of dielectrics or conductive abrasive powders, which negatively affect the environment and operator health in terms of improving product quality and reducing costs. However, with the use and development of environmentally friendly fluids as dielectrics in EDM, reduction in tool, dielectric and machining costs and improvement in product quality (surface quality, dimensional accuracy, etc.) can be achieved.

4.5 ECO-FRIENDLY DIELECTRICS

It has been determined that prolonged contact with the petroleum-based hydrocarbon containing dielectric liquid used in electrical discharge machining causes skin irritation, and harmful gases and aerosols released during processing cause respiratory diseases (Singh et al., 2021). The aerosols released during processing consist of electrode and dielectric components. It has been seen that dielectrics consisting of hydrocarbon compounds help the emission of these gases, so it is recommended to use dielectrics that control the emission of harmful gases and environmental health (Kalyon and Fatatit, 2019). In order to eliminate these negative situations, it is important that the operator works hygienically and carefully, that the ventilation is sufficient to remove harmful gases from the environment in a controlled manner, and that appropriate and environmentally friendly dielectrics are used. Asgar and Singholi mentioned that toxic fumes, vapors, and aerosols released from hydrocarbon-based dielectrics used in EDM cause various health problems (Asgar and Singholi, 2021). Yadav et al. mentioned that CO_2, CO gases and benzene emitted during processing with conventional dielectric fluid are emitted into the environment (Yadav et al., 2021). Jose et al. have revealed by the Hazard and Operability Analysis (HAZOP) method that the kerosene used in the EDM of the casting steel causes the aerosol release of benzene, toluene, ethyl benzene, xylene, polycyclic aromatic hydrocarbon, nickel, and chromium. They also stated that the aerosol density increased with the discharge current (Jose et al., 2010). Some studies mention CO and methane (CH_4) emission in the use of mineral oil in EDM (Chakraborty et al., 2015). Referring to the effect of hydrocarbon compounds as additives, Rehbein et al. suggest the use of harmless compounds for health as additives added to mineral oil (Rehbein et al., 2004). In a review study that mentions the release of aerosols containing polycyclic aromatic hydrocarbon (PAH), benzene and harmful oil minerals from mineral oils, it is mentioned that CO, NO_2 and ozone particles are released in the aerosol as a result of the use of water-based dielectric (Baroi et al., 2022). Mineral oils are often used in EDM processes, especially for dimensional accuracy and minimal tool wear. However, there are strict guidelines in the ISO 14000 standard that cover operators' health and safety issues, production cost, material and energy consumption, waste management, and sustainable production and green environmental criteria such as the environmental impact of the operation (Chaudhury and Samantaray, 2021). Therefore, it is recommended to use synthetic fluids instead. However, synthetic fluids' high production cost and low environmental impact limit their use except for mass production (Valaki and Rathod, 2016). In addition, dielectrics used due to the content of dangerous substances such as chromium, lead, and cadmium (depending on the material being processed and tool material) cause environmental problems (water and soil pollution) as waste, and special and expensive procedures are required to neutralize them (Gumpu et al., 2015; Radu et al., 2020). For these reasons, using non-renewable mineral oil as an EDM insulator seems inappropriate. Recent research on the EDM process focused on finding solutions that will make it more environmentally friendly and less harmful to the operator. Such a solution consists of using eco-friendly oils fluids instead of hydrocarbon-based or synthetic oils without affecting process performance in terms of quality and productivity (Singaravel et al., 2020). Liquid and gas dielectrics, called eco-friendly or green; it can be categorized as various gases that do not spread to the environment as waste and show low greenhouse gas emissions, and biodegradable vegetable oils. Dry EDM, in which various gases are used as dielectric, is considered an economical and eco-friendly method without complex washing mechanisms (Nishant Singh et al., 2016). Here, the pressurized gas is sent via a nozzle or electrode to remove the debris, provide the dielectric medium, and cool it. In order for this process to be sustainable, it is recommended to use environmentally friendly green gases such as air, argon, helium, nitrogen, and oxygen, which are not released into the environment as waste (Nishant Singh et al., 2021). The researchers observed high dimensional accuracy and thin white layer formation in gas EDM processes (Gholipoor et al., 2015). However, it has been reported that hydrodynamic forces do not influence the plasma channel as in liquid dielectrics and the continuity of expansion due to the low density of gas dielectrics (Govindan and Joshi, 2010). It has been emphasized that the debris cannot be removed at high discharges; sticking to the electrode occurs and, therefore, decreases in MRR and surface quality (Dhakar and Dvivedi, 2016; Liqing and Yingjie, 2013). On the other hand, it can be said that the debris globules shrink under high pressure and tension, making it easier to transport with the pressurized dielectric medium (Pragadish and Kumar, 2016; Saha and Choudhury, 2009). In addition, to eliminate the disadvantages of dry EDM, the near-dry EDM technique can be used by mixing compressed gas and minimum liquid dielectrics (Nishant Kumar and Singh, 2015).

Natural esters obtained from different plant seed oils are seen as innovative insulating materials (dielectrics) that can be used instead of petroleum-based oils with their high viscosity and flash point. Some researchers have suggested converting vegetable oils to biodiesel and using low-viscosity biodiesel as dielectric in EDM. Some researchers have suggested converting vegetable oils (trans-esterified) into biodiesel and using low viscosity biodiesel as a dielectric in EDM (Kong and Yeo, 2017; Yunus Khan et al., 2020). For example, the viscosity of jatropha, 62.54 cSt, and rice bran oil, 89.82 cSt, was reduced to 4.76 and 8.92 cSt, respectively, by the transesterification process (Chakraborty et al., 2022). Its ability to absorb 20–30 times more moisture than petroleum-based oils increases the availability of natural ester dielectrics against harmful gases. In addition to its environmental and thermal advantages, the degradation of these dielectrics due to partial discharges may increase when used as insulators (Mohamed et al., 2017). Also, natural esters tend to degrade by oxidation due to unsaturation. Naturally, the preservation of saturation and fat poses a problem in terms of cost. However, due to environmental effects, biodegradability problems, and limited reserve resources of petroleum-derived oils, green dielectrics need to be investigated, overcome, and developed (Rafiq et al., 2020). In general, high MRR is obtained in vegetable oils, because the heat energy released with high thermal conductivity accelerates the melting-evaporation of the material. Due to the high viscosity of vegetable oils, it increases the energy density by narrowing the plasma channel and increasing the washing pressure according to the dielectric level (Das et al., 2019). This creates deep craters and dense resolidified regions while improving the MRR. Conversely, the use of vegetable oils in the EDM process has benefits such as low carbon content, reduction of the black layer that prevents tool wear and the density of the white layer that affects the workpiece hardness (Das et al., 2021; Mabbu, 2021). As a renewable resource, vegetable oil is biodegradable, has a high flash point, insulation, and corrosion resistance, opening a new direction in the application of biological media. The technical properties of dielectrics used in research on EDM are listed in Table 4.1.

TABLE 4.1

Technical characteristics of dielectrics

Dielectrics	Density	Kinematic viscosity	Thermal conductivity	Specific heat	Flash point	Fire point	Breakdown voltage	Dielectric constant	Oxygen content
Unit	g/cm³	Cst	W/mK	Kj/kgK	C	C	kV	–	–
Kerosene	0.8	1.21	0.128	2.01	47	52	48	2.12	–
Sunflower seeds	0.92	4.9	0.152	–	330	355	–	3.174	–
Polanga	0.869	14.4	–	–	140	–	32	–	–
Pongamia	0.885	7.15	0.119		102	107	22	3.42	0.52
Jatropha	0.868	6.6	0.146	1.93	171	–	26	3.24	1.1
Canola	0.87	4.7	0.179	1.91	450		24.4	3.28	–
Neem	0.868	6.63	0.126	1.96	109	113	25	3.67	0.62
Coconut	–	–	–	–	–	–	70	–	–
Sunflower	0.88	4.9	–	–	439	–	60	3	–
Rice Bran	0.8627	8.92	–	–	–	–	104	–	–
Olive	0.912	43.60	–	–	177	–	26–41	–	–
Mustard	0.967	63.40	–	–	110	–	28	–	–
Amla	0.966	102.8	–	–	210	–	50	–	–
Palm	0.87	40.24	0.1708	1.902	160	–	51–64	–	–

Carbon-based oils are known to introduce hazardous environmental exposures and poorly machined surface quality. These oils are not biodegradable and cause environmental pollution as waste. The thermal effects that occur with high discharge energy also cause the emission of toxic elements in the dielectric during evaporation (Mishra and Routara, 2020; Valaki et al., 2019). The radicals in vegetable oils increase with oxidation, shortening the oxidative stability and shelf life, and can easily react with different molecules and harm the environment and living things. Antioxidants are chemical compounds that slow the oxidation (increase of free radicals) of fats. The antioxidant, which is also used as an additive, is oxidized and its stability deteriorates over time. In order to prevent degradation, there is always a need for a substance to fill the antioxidant gap in the oil (Bharti and Singh, 2020; Gupta et al., 2022). Dzülkifli et al. proposed a water emulsion of canola oil as the dielectric instead of petroleum-based oils in EDM. The dielectric properties of the canola oil emulsion, namely thermal conductivity (0.143 W/mK), heat capacity (0.376 MJ/m³K), thermal dissipation (0.380 m²/M), pH (8.490) and viscosity (220.200 cp) were measured, and the best blend combination 10% water containing 35% Tween 20 emulsifier with heat applied was found. It was reported that the dielectric's thermal conductivity and thermal diffusion, which was exposed to 75°C for 120 minutes, increased, and its thermal capacity and viscosity decreased (Dzulkifli et al., 2021).

Thanigaiselvan et al. investigated breakdown voltage, viscosity, flash point and flash point performance indicators in the use of rapeseed and pongamia oil with or without additives as insulation materials. The researchers increased the oxidative stability of these oils by using 2.5 or 5 g of six different antioxidant additives (natural and synthetic) per 500 ml of oil. Butylated Hydroxy Anisol (BHA), Butylated Hydroxy Toluene (BHT) and α-Tocopherol Acetate additives were found for the additive oil to have lower carbon emission and improved breakdown voltage and flash point in both oils. BHT, α–T and Propyl Gallate (PG) additives have been suggested for low viscosity with high heat transfer rate (Thanigaiselvan et al., 2015). Kunju and Shemim compared the breakdown voltage (BDT), tan-delta (tanδ) and viscosity of coconut, rice bran, sesame, and sunflower oils with additive mineral oil. In the experimental study, TiO_2, Fe_3O_4, SiO_2, h-BN and Al_2O_3 nanoparticles were added to 550 ml of oil at a ratio of 0.1 g/100 ml. While the addition of SiO_2 nanoparticles to dielectric oils causes a decrease in BDT, rice bran oil with h-BN is recommended as the most ideal insulation material with its high breaking voltage (150–172 kV/cm) and ideal tan-delta (0.18/mm). However, it has been stated that the viscosity of mineral oil (10.48 cSt) is lower than that of rice bran oil (52.55 cSt), and the viscosity of natural oils can be reduced by reducing particle size or using surfactants (Kunju and Shemim, 2019). Considering the environmental health effects of harmful gases, pollutants and wastes arising from electrode materials and dielectric components released during EDM, it is necessary to choose the most suitable processing method (bio-dielectric liquid, water, gas and dry EDM) for the material to be processed and the tool material used. It is also known that emissions and harmful wastes vary according to processing parameters. In this case, it is important to determine the appropriate input parameters for the chosen method before the process. In any case, operators' awareness of EDM risks should be increased, and safety measures should be established that reduce and prevent the release of non-environmentally friendly pollutants.

4.6 RECENT STUDIES ON EDM AND WEDM

This section examines recent research on the use of environmental dielectrics in die-sinking and wire EDM techniques.

4.6.1 DIE-SINKING EDM

Ganachari et al. drilled a hole in spring steel to compare EDM and near-dry EDM using a copper tube electrode. As a dielectric, a mixture of oxygen and water is used in near-dry EDM, and only air is used in dry EDM. With a flow rate of 5 lt/min, they provided a dielectric transition from the copper pipe to the workpiece. It was stated that the current had little effect on the MRR in dry-EDM, while the pressure effect was less for TWR. It has been said that Tension and Ton parameters have no significant effect on MRR for near dry EDM, while pressure and Ton parameters have little effect for TWR. As a result of SEM examinations of the surface morphology, they stated that near-dry EDM was more productive in terms of microcracks, deep craters, white layer, conical and impact holes (Ganachari et al., 2019). Viswanth et al. reviewed dielectric fluids for aerosol and metallic emissions that directly affect the environmental balance in sustainable and environmentally friendly EDM applications. It has been observed that hybrid processing (powder mixed EDM, etc.) will directly benefit environmental factors. Factors such as low viscosity, good thermal conductivity, good flow rate, less white layer formation due to low carbon content increase the preference of these dielectrics due to the environmental sensitivity of dielectrics with vegetable oil, water, air, and powder. The authors emphasized that bio-oils such as palm and peanut oils could be used in future applications from their high flash points (Srinivas et al., 2018). Pellegrini and Ravasio investigated the environmental effects of liquid or gaseous dielectrics on energy consumption, electrode wear, machining performance, and sustainability in drilling using μ-EDM and tungsten and brass electrodes in AISI 316L and Ti6Al4V materials. In the experiments, current in the range of 49–90 A and voltage in the 79–110 V range were used. As gas dielectrics show less effective washing than liquid dielectrics, the drilling rate is decreased. The stainless steel-tungsten tool pair achieved the best performance and environmental impact in water and air. The stainless steel-brass combination determined water or vegetable oil as the most effective dielectric. While water and vegetable oil are effective in the titanium-WC couple, oxygen is an effective dielectric in terms of tool performance, although it is weak in terms of performance. In the titanium-rice couple, water is the most effective dielectric, and vegetable oil significantly affects energy consumption compared to kerosene (Pellegrini and Ravasio, 2020). Singh et al. evaluated the effect of bio-dielectrics on machining performance in the EDM of AlSiC composite. Kerosene and two different vegetable oils based on Pongamia and Jatropha were used as dielectric. In the use of Pongamia and Jatropha oil, 45–75 and 30%–60% improvement in machining speed, 23 and 18% improvement in surface quality, 20–30 and 20%–25% improvement in electrode wear were achieved, respectively. The high heat released during machining accelerated the oxidation of the vegetal dielectric and contributed to the increase in the temperature on the workpiece surface, resulting in an increase in MRR. The authors noted that surfaces treated with vegetable oils had fewer surface cracks, debris particles, and a thin-smooth recast layer than kerosene (Singh et al., 2020). Ming et al. investigated the effects of using sunflower seed oil (SSO) and kerosene as dielectric on machining performance in die-sinking EDM of SKD11 tool steel with different tools (Cu, W, Cu-W and graphite). In the tests, the effects of dielectric and basic processing parameters on MRR, Ra, energy efficiency per volume (EEV) and exhaust emissions characteristics (EEC) were investigated. It was stated that Ra increased due to carbon adhesion to the graphite instrument. While high MRR and low emission rate were obtained in experiments with copper instruments using SSO, low EEV and Ra values were obtained using graphite. SSO and kerosene dielectrics are said to show similar performance characteristics. SSO is a dielectric that can be used instead of kerosene, which does not pollute soil and water resources, is environmentally friendly, biodegradable, renewable and does not pose a fire risk with its high flash point (Ming et al., 2022). Radu et al. stated that vegetable dielectrics were deteriorated when exposed to heating, light and oxidation and investigated the change of physicochemical properties of vegetable oils after use in EDM. The machining time was analyzed in EDM of 17–4 PH, AZ31B magnesium alloy and AA 7075-T7351. They evaluated the refractive index, dynamic viscosity, and spectrum analysis of the dielectrics (mineral, sunflower, and soybean oil) used during processing. It was emphasized that the pre-processing spectrum analysis of vegetable oils showed similar absorption to mineral oil between 190 and 330 nm, and the post-processing spectrum of other oils was the same as sunflower oil. They observed a small increase in viscosity with the processing of stainless steel compared to other soft materials. It has been reported that there is no structural change in vegetable dielectrics under optimum processing conditions, and its use instead of mineral oil has been proven in terms of cost, environment and operator health (Radu et al., 2020). Viswanth et al. Ton (300–500 µs), Toff (30–90 µs), I (9–21 A), reference voltage (40–60 V), and gap (60–180 µm) variables on MRR, EWR and SR were investigated using pongamia and neem oils. Researchers reported that vegetable dielectrics showed higher MRR and EWR, lower SR compared to conventional oil. The effect of delayed melting and evaporation with increasing gap distance caused the formation of small and shallow craters, resulting in a reduction in SR. In addition, the high gap reduces the effect of viscosity, reducing the possibility of partially solidified residual material that cannot be ejected during washing. Compared to conventional oil, they found relatively low crater marks, debris globules, microcracks, and the formation of bubbles on the surface of the material treated with vegetable dielectric (Viswanth et al., 2020). Powder mixed EDM is a method that increases processing efficiency by improving the processing speed and microhardness by adding conductive particles to the dielectric liquid (Abdudeen et al., 2020). Reddy et al. examined the effects of I, Ton, and Toff parameters on MRR, TWR and SR results in EDM of AISI P20 steel. The performance of sunflower and kerosene dielectrics on MRR, TWR and SR was investigated using three types of electrodes (copper, brass and WC-Cu). Further melting and resolidification with the increase in MRR caused a depth difference, resulting in an increase in surface roughness. In sunflower oil, the best MRR and TWR were obtained with the tungsten-copper tool, while the best SR was obtained with the brass tool (Reddy et al., 2019). Chakraborty et al. investigated the effects of I (4–12 A), Ton (20–320 µs) and V (30–70 V) parameters on MRR and Ra in die-sinking EDM of 1% B4C alloy using different dielectrics (EDM oil, trans esterified jatropha and rice bran). In the experimental study with biodiesel jatropha oil, the researchers obtained the best MRR (330 mg/min), Ra (3.39 μm), white layer (123.5 nm) results. The toxic CO and CO_2 gas emission in vegetable dielectrics is lower than that of EDM oil. The authors noted that viscosity was the determining factor for recast layer thickness. They reported that the material processed using EDM oil has larger craters, rougher surfaces, and equal surface hardness than biodiesel oils (Chakraborty et al., 2022). Supawi et al. focused on the effect of machining parameters on white layer thickness and microhardness in EDM of AISI D2 steel with kerosene and refined palm oil. While the authors reported lower layer thickness and microhardness compared to kerosene in experiments with biodielectric. They observed that cracks formed on the material surface in both dielectric types (Supawi et al., 2022). Ishfaq et al. investigated the dimensional accuracy and microstructure changes in EDM of Inconel 600 alloy with Al, brass, Cu and graphite tools by using canola, amla, olive, sunflower, coconut and mustard oil. It is mentioned that at high breakdown voltage or viscosity, the high discharge energy required for thermal breakdown to occur and spark formation will stabilize arc formation while reducing erosion. The dielectrics with the lowest overcut were specified as canola (0.0165 mm) and sunflower oil (0.0380 mm). Noting that the graphite tool with high thermal conductivity (1,950 W/mK) caused the worst dimensional accuracy (0.39 mm). The authors stated that for low overcutting, coconut oil – Al tool (0.0495 mm), canola oil – brass tool, suggested pairs of sunflower oil – copper tool, amla oil – brass tool (0.0725 mm) and mustard oil – brass tool (0.0425 mm) (Ishfaq et al., 2022).

4.6.2 Wire EDM

In wire EDM (WEDM), pure water is generally used as the dielectric due to the minimum effect of tool wear and environmental protection (Bose and Nandi, 2020). Chakraborty et al. focused on the effect of machining parameters (Ton, Toff, V, dielectric level) on MRR, Ra, dielectric consumption by using environmental and conventional dielectrics in the processing of Ti6Al4V alloy with wire PMEDM. Deionized water, kerosene and surfactant (C18H34O6) added deionized water was used as dielectric, and 10 μm long B4C powder was added into the surfactant as an abrasive. More toxic vapor emission during the use of kerosene, the use of deionized water in wire EDM have been suggested due to both efficiency and environmental effects. Although dust and surfactants mixed into the dielectric help form plasma channels, creating a safe working environment is necessary since they cause environmental pollution and health problems (Chakraborty et al., 2020). Velmurugan et al. analyzed the MRR and Ra changes using reference voltage, pulse duration (PW), pulse interval (PI) and flow rate (F) parameters in ND-WEDM of M2-HSS steel with helium supported molybdenum wire. Using the helium-water mixture as a dielectric, the researchers reported that MRR and Ra increased with V, PW and F, and decreased with PI, due to the high temperature in the processing zone. They determined the optimum parameters as 5V25PW45PI20 for MRR and 3V15PW75PI10F for Ra by Taguchi analysis (Velmurugan et al., 2021). Boopathi examined the effects of the parameters of flow rate, discharge current, pulse time and pulse gap on the kerf width (KW), cutting speed (CS) and wire wear rate (WWR) by using water and coconut oil-added steam as dielectric during WEDM of Monel-K500 alloy. The researcher reported that steam with coconut oil added better KW (6.82%), WWR (15.49%) and CS (21.71%) compared to water vapor (Boopathi, 2022). In another study by Boopathi, the effects of air pressure, flow rate, I and Ton were examined on the results of gas emission concentration (GEC), MRR and relative emission rate (RER) in near dry-WEDM of AISI D2 steel using air-water mixture. Lowering high air pressure and other parameters for low GEC; suggested high Ton, discharge current and flow rate for high MRR, and low discharge current and Ton, high air pressure and flow rate for low RER. According to the author, air pressure MRR and GEC seem to be the most influential parameters (Boopathi, 2021). Kumar and Babu, pulse time (1–5 µs), pulse interval (6–10 µd), reference voltage (50–60 V) and wire speed (6–10 m/min) using air-water mixture in ND-WEDM of Monel 400 alloy investigated the effect of parameters on MRR and Ra. The highest MRR was 0.00003256 kg/min (5 µs, 10 µs, 55 V and 8 m/min) and the lowest Ra was 0.98 µm (1 µs, 6 µs, 60 V and 8 m/min) (Kumar and Babu, 2018). Gunasekaran et al. investigated the effects of pressure, flow rate, discharge current and Ton on WWR and MRR during ND-WEDM of Inconel 718 alloy using deionized water mixed air, oxygen and helium gases with the help of a cryogenically cooled wire. The spark density in the cutting zone, which increases with increasing air pressure, Ton and discharge current, increases MRR and WWR, and efficient washing with increasing flow rate improves MRR and WWR. Researchers have obtained 30%–27% higher efficiency in the WWR of helium and oxygen-doped dielectrics compared to the air mixture. The authors, who achieved 7%–3% efficiency in MRR, stated that the best machining quality would be achieved using oxygen-doped dielectric (Gunasekaran et al., 2022). Boopathi et al. focused on the effects of discharge current (2–4 A), pulse duration (20–30 µs), pulse interval (12–28 µs) and reference voltage (75–100 V) parameters on MRR and Ra in oxygen gas dry WEDM of Al6061 + 3% SiC composite. According to the authors, choosing oxygen gas as the insulating medium increases the machining accuracy, while using water reduces the corrosion effect and tool wear. Optimizing the processing conditions, the authors obtained optimal MRR (10.512 mm³/min) and Ra (2.31 µs) results (Boopathi et al., 2012). Dielectric type, composition and viscosity affect the fumes and vapors released during EDM. A more viscous fluid is effective at absorbing a significant portion of smoke and steam with increased dielectric level but has a negative effect on chip transport and plasma channel expansion. The intense energy in the unexpandable plasma channel can cause the formation of deep craters in the surface structure. An increase in crack density can be observed with rapid thermal changes (Valaki and Rathod, 2016).

4.7 CONCLUSIONS AND FUTURE RESEARCH

Despite the numerous advantages of EDM technology, this process has some limitations, such as large energy consumption, slow material removal rate, hazardous emissions, toxic dielectric and slurry formation. Die-sinking EDM and wire EDM techniques can involve high processing times (depending on the complexity of the part and the desired quality) and thus prolonged exposure to light and oxygen, and high temperatures (depending on the material being processed the operating regime applied). EDMs release emissions and waste of harmful substances and affect the environment and health of operators. These harmful entities vary according to the processing process and input parameters. Hydrocarbon-based dielectrics have the worst impact on the environment and operator's health. With the right selection and combination of processing parameters (I, Ton, Toff, dielectric pressure, etc.), reducing the release of toxic aerosols is possible. At this point, the use of vegetable oil and water dielectrics, dry and near-dry methods is accepted as the right step for eco-friendly machining.

It has been reported that dielectric fluids used in EDM should have higher flash point, high biodegradability, higher oxygen content, lower carbon atom chain, higher breakdown voltage, favorable viscosity, lower toxic emissions and lower volatility. Also, dielectric strength and viscosity are the most important factors for high MRR. In this context, the technical feasibility and qualitative performance of vegetable oils as a substitute for hydrocarbon-based dielectrics and synthetic oils in the EDM process is important. From an economic point of view, the use of vegetable oils as a dielectric will reduce the costs of EDM because many plants such as canola and sunflower are available worldwide. In addition, it is mentioned that vegetable oils maintain their dimensional accuracy under roughing conditions and increase in processing speed compared to conventional hydrocarbon oils. This means increased productivity and lower production costs. Since vegetable oils have higher flash and burning points than conventional mineral oil, they offer a low risk of burning, supporting them to be environmentally and operator sensitive. Due to vegetable oils' high viscosity and low shelf life, low surface quality and MRR are not observed in the EDM process. As a solution to this situation, transesterification of pure or used vegetable oils is recommended. In the literature, improvements in the viscosity, shelf life, flash point, density, breakdown voltage and dielectric strength of biodiesel obtained by the transesterification process have been observed in accordance with EDM. Also, water, dry or near-dry EDM techniques perform well over hydrocarbon-based dielectric machining in terms of EWR, MRR, TWR and integrity of machined surfaces. Advances in machined material technologies require the perfect interaction of numerous EDM machining parameters such as workpiece-tool material and geometry, dielectric specifications, voltage, spark-on time, spark-off time, wire feed speed, wire tension. At the same time, minimum tool wear, low dielectric consumption and energy consumption, minimization of wastes and gases harmful to the operator and the environment are very important for sustainability. In this context, it is challenging to reach the best outputs. It is useful to benefit from multi-purpose optimization and analysis techniques to obtain optimum processing parameters. Future research in this context can significantly contribute to sustainability by supporting research at minimum cost. Green dielectrics should be included more in research for sustainable and eco-friendly production. In addition, the effect of different vegetable dielectrics converted to biodiesel on the processing performance should be examined in detail.

REFERENCES

Abdudeen, Asarudheen, Jaber E. Abu Qudeiri, Ansar Kareem, Thanveer Ahammed, and Aiman Ziout. 2020. 'Recent Advances and Perceptive Insights into Powder-Mixed Dielectric Fluid of EDM'. *Micromachines* 11 (8). Multidisciplinary Digital Publishing Institute: 754. doi:10.3390/mi11080754.

Almeida, Sergio Tadeu, John Mo, Cees Bil, Songlin Ding, and Xiangzhi Wang. 2021. 'Conceptual Design of a High-Speed Wire EDM Robotic End-Effector Based on a Systematic Review Followed by TRIZ'. *Machines* 9 (7). Multidisciplinary Digital Publishing Institute: 132. doi:10.3390/machines9070132.

Akıncıoğlu, Sıtkı. 2022. 'Taguchi Optimization of Multiple Performance Characteristics in the Electrical Discharge Machining of the Tigr2'. *Facta Universitatis, Series: Mechanical Engineering (FU Mech Eng)* 20 (2): 237–253. doi:10.22190/FUME201230028A.

Asgar, Md Ehsan, and Ajay Kumar Singh Singholi. 2018. 'Parameter Study and Optimization of WEDM Process: A Review'. *IOP Conference Series: Materials Science and Engineering* 404 (September). IOP Publishing: 012007. doi:10.1088/1757-899X/404/1/012007.

Asgar, Md Ehsan, and Ajay Kumar Singh Singholi. 2021. 'Study of the Effect of Dielectric on Performance Measure in EDM'. In *Advances in Industrial and Production Engineering*, edited by Rakesh Kumar Phanden, K. Mathiyazhagan, Ravinder Kumar, and J. Paulo Davim, 843–850. Mechanical Engineering. Singapore: Springer. doi:10.1007/978-981-33-4320-7_75.

Baroi, Binoy Kumar, Jagadish, and Promod Kumar Patowari. 2022. 'A Review on Sustainability, Health, and Safety Issues of Electrical Discharge Machining'. *Journal of the Brazilian Society of Mechanical Sciences and Engineering* 44 (2): 59. doi:10.1007/s40430-021-03351-4.

Bharti, Rupam, and Bhaskar Singh. 2020. 'Green Tea (Camellia Assamica) Extract as an Antioxidant Additive to Enhance the Oxidation Stability of Biodiesel Synthesized from Waste Cooking Oil'. *Fuel* 262 (February): 116658. doi:10.1016/j.fuel.2019.116658.

Bisaria, Himanshu, and Pragya Shandilya. 2015. 'Machining of Metal Matrix Composites by EDM and Its Variants: A Review.' In *DAAAM International Scientific Book*, 14: 267–282. 23. Vienna, Austria: DAAAM International. doi:10.2507/daaam.scibook.2015.23.

Boopathi, S., K. Sivakumar, and R. Kalidas. 2012. 'Parametric Study of Dry WEDM Using Taguchi Method'. *International Journal of Engineering Research and Development (ISSN: 2278-800X)* 2 (4). Citeseer: 1–6.

Boopathi, Sampath. 2021. 'An Investigation on Gas Emission Concentration and Relative Emission Rate of the Near-Dry Wire-Cut Electrical Discharge Machining Process'. *Environmental Science and Pollution Research*, November. doi:10.1007/s11356-021-17658-1.

Boopathi, Sampath. 2022. 'Performance Improvement of Eco-Friendly Near-Dry Wire-Cut Electrical Discharge Machining Process Using Coconut Oil-Mist Dielectric Fluid'. *Journal of Advanced Manufacturing Systems*, June. World Scientific Publishing Co., 1–20. doi:10.1142/S0219686723500178.

Bose, Soutrik, and Titas Nandi. 2020. 'A Novel Optimization Algorithm on Surface Roughness of WEDM on Titanium Hybrid Composite'. *Sādhanā* 45 (1): 236. doi:10.1007/s12046-020-01472-5.

Çakıroğlu, Ramazan. 2022. 'Analysis of EDM Machining Parameters for Keyway on Ti-6Al-4V Alloy and Modelling by Artificial Neural Network and Regression Analysis Methods'. *Sādhanā* 47 (3): 150. doi:10.1007/s12046-022-01926-y.

Chakraborty, S., V. Dey, and S. K. Ghosh. 2015. 'A Review on the Use of Dielectric Fluids and Their Effects in Electrical Discharge Machining Characteristics'. *Precision Engineering* 40 (April): 1–6. doi:10.1016/j.precisioneng.2014.11.003.

Chakraborty, S., S. Mitra, and D. Bose. 2020. 'Performance Analysis on Eco-Friendly Machining of Ti6Al4V Using Powder Mixed with Different Dielectrics in WEDM'. *International Journal of Automotive and Mechanical Engineering* 17 (3): 8128–8139. doi: 10.15282/ijame.17.3.2020.06.0610.

Chakraborty, Tapas, Deepti Ranjan Sahu, Amitava Mandal, and Bappa Acherjee. 2022. 'Feasibility of Jatropha and Rice Bran Vegetable Oils as Sustainable EDM Dielectrics'. *Materials and Manufacturing Processes* 38 (1). Taylor & Francis: 1–14. doi:10.1080/10426914.2022.2089891.

Chaudhury, Pallavi, and Sikata Samantaray. 2021. 'A Comparative Study of Different Dielectric Medium for Sustainable EDM of Non-Conductive Material by Electro-Thermal Modelling'. *Materials Today: Proceedings*, International Conference on Recent Advances in Mechanical Engineering Research and Development (ICRAMERD-20), 41 (January): 437–444. doi:10.1016/j.matpr.2020.10.162.

Das, Shirsendu, Swarup Paul, and Biswanath Doloi. 2019. 'An Experimental and Computational Study on the Feasibility of Bio-Dielectrics for Sustainable Electrical Discharge Machining'. *Journal of Manufacturing Processes* 41 (May): 284–296. doi:10.1016/j.jmapro.2019.04.005.

Das, Shirsendu, Swarup Paul, and Biswanath Doloi. 2021. 'Assessment of the Impacts of Bio-Dielectrics on the Textural Features and Recast-Layers of EDM-Surfaces'. *Materials and Manufacturing Processes* 36 (2). Taylor & Francis: 245–255. doi:10.1080/10426914.2020.1832678.

Deshmukh, Suhas P., Ramakant Shrivastava, and Chetan M. Thakar. 2022. 'Machining of Composite Materials through Advance Machining Process'. *Materials Today: Proceedings*, International Conference on Smart and Sustainable Developments in Materials, Manufacturing and Energy Engineering, 52 (January): 1078–1081. doi: 10.1016/j.matpr.2021.10.495.

Dhakar, Krishnakant, Kuldeep Chaudhary, Akshay Dvivedi, and Omkar Bembalge. 2019. 'An Environment-Friendly and Sustainable Machining Method: Near-Dry EDM'. *Materials and Manufacturing Processes* 34 (12). Taylor & Francis: 1307–1315. doi:10.1080/10426914.2019.1643471.

Dhakar, Krishnakant, and Akshay Dvivedi. 2016. 'Parametric Evaluation on Near-Dry Electric Discharge Machining'. *Materials and Manufacturing Processes* 31 (4). Taylor & Francis: 413–421. doi:10.1080/10426914.2015.1037905.

Dzulkifli, Norazkifni Faizura, Azuddin Mamat, and Imtiaz Ahmed Choudhury. 2021. 'The Potential of Water-In-Oil Emulsion of Canola Oil as Dielectric Fluid for EDM Process'. *Archives Of Academia Baru Articles* 72 (2): 129–141.

Equbal, Azhar, and Anoop Kumar Sood. 2014. 'Electrical Discharge Machining: An Overview on Various Areas of Research'. *Manufacturing and Industrial Engineering* 13 (September): 1–6. doi:10.12776/mie.v13i1-2.339.

Ganachari, V. S., U. N. Chate, L. Y. Waghmode, S. A. Mullya, S. S. Shirguppikar, M. M. Salgar, and V. T. Kumbhar. 2019. 'A Comparative Performance Study of Dry and Near Dry EDM Processes in Machining of Spring Steel Material'. *Materials Today: Proceedings*, 9th International Conference of Materials Processing and Characterization, ICMPC-2019, 18 (January): 5247–5257. doi:10.1016/j.matpr.2019.07.525.

Garg, R. K., K. K. Singh, Anish Sachdeva, Vishal S. Sharma, Kuldeep Ojha, and Sharanjit Singh. 2010. 'Review of Research Work in Sinking EDM and WEDM on Metal Matrix Composite Materials'. *The International Journal of Advanced Manufacturing Technology* 50 (5): 611–624. doi:10.1007/s00170-010-2534-5.

Gholipoor, Ahad, Hamid Baseri, and Mohammad Reza Shabgard. 2015. 'Investigation of near Dry EDM Compared with Wet and Dry EDM Processes'. *Journal of Mechanical Science and Technology* 29 (5): 2213–2218. doi:10.1007/s12206-015-0441-2.

Govindan, P., and Suhas S. Joshi. 2010. 'Experimental Characterization of Material Removal in Dry Electrical Discharge Drilling'. *International Journal of Machine Tools and Manufacture* 50 (5): 431–443. doi:10.1016/j.ijmachtools.2010.02.004.

Goyal, Ashish. 2017. 'Investigation of Material Removal Rate and Surface Roughness during Wire Electrical Discharge Machining (WEDM) of Inconel 625 Super Alloy by Cryogenic Treated Tool Electrode'. *Journal of King Saud University – Science, SI: Smart Materials & Applications of New Materials* 29 (4): 528–535. doi:10.1016/j.jksus.2017.06.005.

Groover, Mikell P. 2010. *Fundamentals of Modern Manufacturing*. 4th ed. Hoboken, NJ: John Wiley & Sons. ISBN: 978-0470-467008.

Gumpu, Manju Bhargavi, Swaminathan Sethuraman, Uma Maheswari Krishnan, and John Bosco Balaguru Rayappan. 2015. 'A Review on Detection of Heavy Metal Ions in Water – An Electrochemical Approach'. *Sensors and Actuators B: Chemical* 213 (July): 515–533. doi:10.1016/j.snb.2015.02.122.

Gunasekaran, K., Sampath Boopathi, and M. Sureshkumar. 2022. 'Analysis of a Cryogenically Cooled Near-Dry Wedm Process Using Different Dielectrics'. *Materials and Technology* 56 (2): 179–186. doi:10.17222/mit.2022.397.

Gupta, Akash, and Harish Kumar. 2021. 'Optimization of EDM Process Parameters: A Review of Technique, Process, and Outcome'. In *Advances in Manufacturing and Industrial Engineering*, edited by Ranganath M. Singari, Kaliyan Mathiyazhagan, and Harish Kumar, 981–996. Lecture Notes in Mechanical Engineering. Singapore: Springer. doi: 10.1007/978-981-15-8542-5_87.

Gupta, K., N. K. Jain, and R. F. Laubscher. 2015. 'Spark-Erosion Machining of Miniature Gears: A Critical Review'. *The International Journal of Advanced Manufacturing Technology* 80 (9–12): 1863–1877. doi:10.1007/s00170-015-7130-2.

Gupta, K., and M. K. Gupta. 2019. 'Developments in Non-Conventional Machining for Sustainable Production-A State of Art Review'. *Proceedings of the IMechE, Part C: Journal of Mechanical Engineering Science*, 233 (12): 4213–4232. doi:10.1177/0954406218811982.

Gupta, Vivudh, Pawandeep Singh, Balbir Singh, and R. K. Mishra. 2022. 'Vegetable Oil Based Dielectric Fluids for Electrical Discharge Machining Process: Advancements and Challenges'. *Materials Today: Proceedings*, International Conference on Materials, Processing & Characterization (13th ICMPC), 62 (January): 3129–3132. doi:10.1016/j.matpr.2022.03.395.

Ho, K. H., S. T. Newman, S. Rahimifard, and R. D. Allen. 2004. 'State of the Art in Wire Electrical Discharge Machining (WEDM)'. *International Journal of Machine Tools and Manufacture* 44 (12): 1247–1259. doi:10.1016/j.ijmachtools.2004.04.017.

Ishfaq, Kashif, Muhammad Asad Maqsood, Saqib Anwar, Abdullah Alfaify, and Abdul Wasy Zia. 2022. 'Analyzing Micromachining Errors in EDM of Inconel 600 Using Various Biodegradable Dielectrics'. *Journal of the Brazilian Society of Mechanical Sciences and Engineering* 44 (6): 249. doi:10.1007/s40430-022-03560-5.

Jahan, Muhammad P. 2019. 'Electro-Discharge Machining (EDM)'. In *Modern Manufacturing Processes*, edited by Muammer Koç, Tuğrul Özel, 377–409. Hoboken, NJ: John Wiley & Sons, Ltd. doi:10.1002/9781119120384.ch16.

Jameson, Elman C. 2001. *Electrical Discharge Machining*. Dearborn, MI: Society of Manufacturing Engineers.

Jose, Mathew, S. P. Sivapirakasam, and M. Surianarayanan. 2010. 'Analysis of Aerosol Emission and Hazard Evaluation of Electrical Discharge Machining (EDM) Process'. *Industrial Health* 48 (4): 478–486. doi:10.2486/indhealth.MS1127.

Kalyon, Ali, and Abubaker Yousef Fatatit. 2019. 'The Environmental Impact of Electric Discharge Machining'. *International Journal of Engineering Science and Application* 3 (3): 123–129.

Khilji, I. A., S. N. B. M. S. Saffe, S. Pathak, C. R. Chilakamarry, A. S. B. A. Sani, and V. J. Reddy. 2022. 'Facile Manufacture of Oxide-Free Cu Particles Coated with Oleic Acid by Electrical Discharge Machining'. *Micromachines*, 13(6), 969.

Koyano, Tomohiro, Shodai Suzuki, Akira Hosokawa, and Tatsuaki Furumoto. 2016. 'Study on the Effect of External Hydrostatic Pressure on Electrical Discharge Machining'. *Procedia CIRP*, 18th CIRP Conference on Electro Physical and Chemical Machining (ISEM XVIII), 42 (January): 46–50. doi:10.1016/j.procir.2016.02.184.

Kumar, Anish, Dr Vinod Kumar, and Dr Jatinder Kumar. 2011. 'A Review on the State of the Art in Wire Electric Discharge Machining (Wedm) Process'. *SSRN Scholarly Paper*. Rochester, NY. https://papers.ssrn.com/abstract=3498784.

Kumar, K., L. Senthil, and S. K. Dhinesh. 2020. 'Dielectric Fluid Parameter Optimization in Machining of Composite Material Using WEDM Process'. *IOP Conference Series: Materials Science and Engineering* 995 (1). IOP Publishing: 012016. doi:10.1088/1757-899X/995/1/012016.

Kumar, N., E. Arun, and A. Suresh Babu. 2018. 'Influence of Input Parameters on the Near-Dry WEDM of Monel Alloy'. *Materials and Manufacturing Processes* 33 (1). Taylor & Francis: 85–92. doi:10.1080/10426914.2017.1279297.

Kumar, Sanjeev, Rupinder Singh, T. P. Singh, and B. L. Sethi. 2009. 'Surface Modification by Electrical Discharge Machining: A Review'. *Journal of Materials Processing Technology* 209 (8): 3675–3687. doi:10.1016/j.jmatprotec.2008.09.032.

Kunju, K. Bijuna, and S. S. Shemim. 2019. 'Dielectric Properties of Eco-Friendly Nanofluids'. *Journal of Physics: Conference Series* 1355 (1). IOP Publishing: 012034. doi:10.1088/1742-6596/1355/1/012034.

Leão, Fábio N., and Ian R. Pashby. 2004. 'A Review on the Use of Environmentally-Friendly Dielectric Fluids in Electrical Discharge Machining'. *Journal of Materials Processing Technology*, 14th Interntaional Symposium on Electromachining (ISEM XIV), 149 (1): 341–346. doi:10.1016/j.jmatprotec.2003.10.043.

Leppert, Tadeusz. 2018. 'A Review on Ecological and Health Impacts of Electro Discharge Machining (EDM)'. *AIP Conference Proceedings* 2017 (1). American Institute of Physics: 020014. doi:10.1063/1.5056277.

Liqing, L., and S. Yingjie. 2013. 'Study of Dry EDM with Oxygen-Mixed and Cryogenic Cooling Approaches'. *Procedia CIRP*, Proceedings of the 17th CIRP Conference on Electro Physical and Chemical Machining (ISEM), 6 (January): 344–350. doi:10.1016/j.procir.2013.03.055.

Mabbu, Dastagiri, Srinivasa Rao P., and Madar Valli P. 2021. 'Experimental Investigation of EDM Process Parameters by Using Pongamia Pinnata Osil Blends as Dielectric Medium'. *Journal of Computational & Applied Research in Mechanical Engineering (JCARME)* 11 (1). Shahid Rajaee Teacher Training University (SRTTU): 47–56. doi:10.22061/jcarme.2019.5056.1620.

Marashi, Houriyeh, Davoud M. Jafarlou, Ahmed A. D. Sarhan, and Mohd Hamdi. 2016. 'State of the Art in Powder Mixed Dielectric for EDM Applications'. *Precision Engineering* 46 (October): 11–33. doi:10.1016/j.precisioneng.2016.05.010.

Ming, Wuyi, Zhuobin Xie, Chen Cao, Mei Liu, Fei Zhang, Yuan Yang, Shengfei Zhang, Peiyan Sun, and Xudong Guo. 2022. 'Research on EDM Performance of Renewable Dielectrics under Different Electrodes for Machining SKD11'. *Crystals* 12 (2). Multidisciplinary Digital Publishing Institute: 291. doi:10.3390/cryst12020291.

Mishra, B. P., and B. C. Routara. 2020. 'Evaluation of Technical Feasibility and Environmental Impact of Calophyllum Inophyllum (Polanga) Oil Based Bio-Dielectric Fluid for Green EDM'. *Measurement* 159 (July): 107744. doi:10.1016/j.measurement.2020.107744.

Mohamed, Mahmoud, Muhammad Amirul Hafiz Bin Bakharazi, Hiroshi Kitagawa, Satoshi Matsumoto, and Masamichi Kato. 2017. 'Partial Discharge Inception Voltage Measurements of Ester Dielectric Fluid for Insulation Diagnosis'. In *2017 6th International Youth Conference on Energy (IYCE)*, 1–4. doi:10.1109/IYCE.2017.8003705.

Mohd Abbas, Norliana, Darius G. Solomon, and Md. Fuad Bahari. 2007. 'A Review on Current Research Trends in Electrical Discharge Machining (EDM)'. *International Journal of Machine Tools and Manufacture* 47 (7): 1214–1228. doi:10.1016/j.ijmachtools.2006.08.026.

Ng, P. S., S. A. Kong, and S. H. Yeo. 2017. 'Investigation of Biodiesel Dielectric in Sustainable Electrical Discharge Machining'. *The International Journal of Advanced Manufacturing Technology* 90 (9): 2549–2556. doi:10.1007/s00170-016-9572-6.

Niamat, Misbah, Shoaib Sarfraz, Haris Aziz, Mirza Jahanzaib, Essam Shehab, Wasim Ahmad, and Salman Hussain. 2017. 'Effect of Different Dielectrics on Material Removal Rate, Electrode Wear Rate and Microstructures in EDM'. *Procedia CIRP*,

Complex Systems Engineering and Development Proceedings of the 27th CIRP Design Conference Cranfield University, UK 10–12 May 2017, 60 (January): 2–7. doi:10.1016/j.procir.2017.02.023.

Pei, Jingyu, Lenan Zhang, Jianyi Du, Xiaoshun Zhuang, Zhaowei Zhou, Shunkun Wu, and Yetian Zhu. 2017. 'A Model of Tool Wear in Electrical Discharge Machining Process Based on Electromagnetic Theory'. *International Journal of Machine Tools and Manufacture* 117 (June): 31–41. doi:10.1016/j.ijmachtools.2017.03.001.

Pellegrini, Giuseppe, and Chiara Ravasio. 2020. 'A Sustainability Index for the Micro-EDM Drilling Process'. *Journal of Cleaner Production* 247 (February): 119136. doi:10.1016/j. jclepro.2019.119136.

Pragadish, N., and M. P. Kumar. 2016. 'Optimization of Dry EDM Process Parameters Using Grey Relational Analysis'. *Arabian Journal for Science and Engineering* 41 (11): 4383–4390. doi:10.1007/s13369-016-2130-6.

Prakash, Ved, Shubham, P. Kumar, P. K. Singh, A. K. Das, S. Chattopadhyaya, A. Mandal, and A. R. Dixit. 2018. 'Surface Alloying of Miniature Components by Micro-Electrical Discharge Process'. *Materials and Manufacturing Processes* 33 (10). Taylor & Francis: 1051–1061. doi:10.1080/10426914.2017.1364755.

Qudeiri, Jaber E. Abu, Aiman Zaiout, Abdel-Hamid I. Mourad, Mustufa Haider Abidi, and Ahmed Elkaseer. 2020. 'Principles and Characteristics of Different EDM Processes in Machining Tool and Die Steels'. *Applied Sciences* 10 (6). Multidisciplinary Digital Publishing Institute: 2082. doi:10.3390/app10062082.

Radu, Maria-Crina, Raluca Tampu, Valentin Nedeff, Oana-Irina Patriciu, Carol Schnakovszky, and Eugen Herghelegiu. 2020. 'Experimental Investigation of Stability of Vegetable Oils Used as Dielectric Fluids for Electrical Discharge Machining'. *Processes* 8 (9). Multidisciplinary Digital Publishing Institute: 1187. doi:10.3390/ pr8091187.

Rafiq, Muhammad, Muhammad Shafique, Anam Azam, Muhammad Ateeq, Israr Ahmad Khan, and Abid Hussain. 2020. 'Sustainable, Renewable and Environmental-Friendly Insulation Systems for High Voltages Applications'. *Molecules* 25 (17). Multidisciplinary Digital Publishing Institute: 3901. doi:10.3390/molecules25173901.

Ramasawmy, H., and L. Blunt. 2004. 'Effect of EDM Process Parameters on 3D Surface Topography'. *Journal of Materials Processing Technology* 148 (2): 155–164. doi:10.1016/ S0924-0136(03)00652-6.

Razzell, A. G. 2000. 'Joining and Machining of Ceramic Matrix Composites'. In *Comprehensive Composite Materials*, 689–697. Amsterdam: Elsevier. doi:10.1016/B0-08-042993-9/00113-3.

Reddy, G. Gowtham, Balasubramaniyan Singaravel, and K. Chandra Shekar. 2019. 'Experimental Investigation of Sunflower Oil as Dielectric Fluid in Die Sinking Electric Discharge Machining Process'. *Materials Science Forum* 969. Trans Tech Publications Ltd: 715–719. doi:10.4028/www.scientific.net/MSF.969.715.

Reddy, M. Chaitanya, K. Venkata Rao, and Gamini Suresh. 2021. 'An Experimental Investigation and Optimization of Energy Consumption and Surface Defects in Wire Cut Electric Discharge Machining'. *Journal of Alloys and Compounds* 861 (April): 158582. doi:10.1016/j.jallcom.2020.158582.

Rehbein, W., H.-P. Schulze, K. Mecke, G. Wollenberg, and M. Storr. 2004. 'Influence of Selected Groups of Additives on Breakdown in EDM Sinking'. *Journal of Materials Processing Technology*, 14th Interntaional Symposium on Electromachining (ISEM XIV), 149 (1): 58–64. doi:10.1016/j.jmatprotec.2004.02.029.

Rohith, R., B. K. Shreyas, S. Kartikgeyan, B. A. Sachin, K. Umesha, and T. S. Nanjundeswaraswamy. 2019. 'Selection of Non-Traditional Machining Process'. *International Journal of Engineering Research & Technology* 8 (11). IJERT-International Journal of Engineering Research & Technology: 148–155.

Saha, Sourabh K., and S. K. Choudhury. 2009. 'Experimental Investigation and Empirical Modeling of the Dry Electric Discharge Machining Process'. *International Journal of Machine Tools and Manufacture* 49 (3): 297–308. doi:10.1016/j.ijmachtools.2008.10.012.

Shirguppikar, Shailesh S., and Uday A. Dabade. 2018. 'Experimental Investigation of Dry Electric Discharge Machining (Dry EDM) Process on Bright Mild Steel'. *Materials Today: Proceedings*, International Conference on Emerging Trends in Materials and Manufacturing Engineering (IMME17), March 10–12, 2017, 5 (2, Part 2): 7595–7603. doi:10.1016/j.matpr.2017.11.432.

Singaravel, B., K. Chandra Shekar, G. Gowtham Reddy, and S. Deva Prasad. 2020. 'Experimental Investigation of Vegetable Oil as Dielectric Fluid in Electric Discharge Machining of Ti-6Al-4V'. *Ain Shams Engineering Journal* 11 (1): 143–147. doi:10.1016/j.asej.2019.07.010.

Singh, Nishant K., Pulak M. Pandey, K. K. Singh, and Manish K. Sharma. 2016. 'Steps towards Green Manufacturing through EDM Process: A Review'. Edited by Shashi Dubey. *Cogent Engineering* 3 (1). Cogent OA: 1272662. doi:10.1080/23311916.2016.1272662.

Singh, Nishant K., Yashvir Singh, Abhishek Sharma, Amneesh Singla, and Prateek Negi. 2021. 'An Environmental-Friendly Electrical Discharge Machining Using Different Sustainable Techniques: A Review'. *Advances in Materials and Processing Technologies* 7 (4). Taylor & Francis: 537–566. doi:10.1080/2374068X.2020.1785210.

Singh, Nishant K., Lalit Yadav, and Sachin Lal. 2020. 'Experimental Investigation for Sustainable Electric Discharge Machining with Pongamia and Jatropha as Dielectric Medium'. *Advances in Materials and Processing Technologies* 8 (2). Taylor & Francis: 1–20. doi:10.1080/2374068X.2020.1860499.

Singh, Nishant Kumar, and K. K. Singh. 2015. 'Review on Recent Development in Environmental-Friendly EDM Techniques'. *Advances in Manufacturing Science and Technology* 39 (1).

Singh, Ramandeep, Neetesh Kumar Sah, and Varun Sharma. 2021. 'Development and Characterization of Unitary and Hybrid Al2O3 and ZrO Dispersed Jatropha Oil-Based Nanofluid for Cleaner Production'. *Journal of Cleaner Production* 317 (October): 128365. doi:10.1016/j.jclepro.2021.128365.

Srinivas Viswanth, V., R. Ramanujam, and G. Rajyalakshmi. 2018. 'A Review of Research Scope on Sustainable and Eco-Friendly Electrical Discharge Machining (E-EDM)'. *Materials Today: Proceedings*, International Conference on Materials Manufacturing and Modelling, ICMMM – 2017, 9–11, March 2017, 5 (5, Part 2): 12525–12533. doi:10.1016/j.matpr.2018.02.234.

Supawi, Aiman, Said Ahmad, Nurul Farahin Mohd Joharudin, and Fazida Karim. 2022. 'Surface Integrity of RBD Palm Oil as a Bio Degradable Oil Based Dielectric Fluid on Sustainable Electrical Discharge Machining (EDM) of AISI D2 Steel'. *Transdisciplinary Research and Education Center for Green Technologies* 9 (1): 41–48. doi:10.5109/4774215.

Thanigaiselvan, R., T. Sree Renga Raja, and R. Karthik. 2015. 'Investigations on Eco Friendly Insulating Fluids from Rapeseed and Pongamia Pinnata Oils for Power Transformer Applications'. *Journal of Electrical Engineering and Technology* 10 (6). The Korean Institute of Electrical Engineers: 2348–2355. doi:10.5370/JEET.2015.10.6.2348.

Valaki, Janak B., and Pravin P. Rathod. 2016. 'Assessment of Operational Feasibility of Waste Vegetable Oil Based Bio-Dielectric Fluid for Sustainable Electric Discharge Machining (EDM)'. *The International Journal of Advanced Manufacturing Technology* 87 (5): 1509–1518. doi:10.1007/s00170-015-7169-0.

Valaki, Janak B., Pravin P. Rathod, and C. D. Sankhavara. 2016. 'Investigations on Technical Feasibility of Jatropha Curcas Oil Based Bio Dielectric Fluid for Sustainable Electric Discharge Machining (EDM)'. *Journal of Manufacturing Processes* 22 (April): 151–160. doi:10.1016/j.jmapro.2016.03.004.

Valaki, Janak B., Pravin P. Rathod, and Ajay M. Sidpara. 2019. 'Sustainability Issues in Electric Discharge Machining'. In *Innovations in Manufacturing for Sustainability*, edited by Kapil Gupta, 53–75. Şekillendirme, İşleme ve Triboloji. Cham: Springer International Publishing. doi:10.1007/978-3-030-03276-0_3.

Velmurugan, D., T. Yuvaraj, V. Raguraman, B. Sampath, and M. Sureshkumar. 2021. 'Experimental Investigations on Eco-Friendly Helium-Mist Near-Dry Wire-Cut EDM of M2-HSS Material'. *Materials Research Proceedings* 19. Materials Research Forum LLC: 175–180.

Viswanth, V. Srinivas, R. Ramanujam, and G. Rajyalakshmi. 2020. 'Improving Productivity with Eco-Friendly Dielectrics in Sustainable EDM Machining of AISI 2507 Super Duplex Stainless Steel'. *International Journal of Precision Technology* 9 (2–3). Inderscience Publishers: 130–151. doi:10.1504/IJPTECH.2020.112061.

Yadav, Avinash, Yashvir Singh, Satyendra Singh, and Prateek Negi. 2021. 'Sustainability of Vegetable Oil Based Bio-Diesel as Dielectric Fluid during EDM Process – A Review'. *Materials Today: Proceedings*, International Conference on Technological Advancements in Materials Science and Manufacturing, 46 (January): 11155–11158. doi:10.1016/j.matpr.2021.01.967.

Yunus Khan, Mohd., P. Sudhakar Rao, and B. S. Pabla. 2020. 'Investigations on the Feasibility of Jatropha Curcas Oil Based Biodiesel for Sustainable Dielectric Fluid in EDM Process'. *Materials Today: Proceedings*, 10th International Conference of Materials Processing and Characterization, 26 (January): 335–340. doi:10.1016/j.matpr.2019.11.325.

Zia, Muhammad Kashif, Salman Pervaiz, Saqib Anwar, and Wael A. Samad. 2019. 'Reviewing Sustainability Interpretation of Electrical Discharge Machining Process Using Triple Bottom Line Approach'. *International Journal of Precision Engineering and Manufacturing-Green Technology* 6 (5): 931–945. doi:10.1007/s40684-019-00043-2.

5 Green Machining of Ceramics

Ergün Ekici and Şenol Bayraktar

5.1 INTRODUCTION

Ceramic materials (CMs) have remarkable properties such as chemical inertness, high strength and high hardness, and low density, especially at very high temperatures (Chawla 2013). On the other hand, engineering ceramics are classified as engineering materials that are difficult to process, due to their high hardness and brittleness. CMs are ideal for wear and high-temperature applications compared to metal and polymer materials. It has become preferred not only in the tool industry but also in the fields of aerospace, defense, and biomedicine in terms of high hardness, strength, low thermal conductivity, biocompatibility, and superior chemical stability properties (Ahlhelm et al. 2016; Ming et al. 2020). Studies on the use of advanced ceramic materials in the area of engineering are still continuing. Silicon nitride ceramic high-speed cutting tools account for 1/3 of the number of ceramic cutting tools in industrial applications (Laouissi et al. 2019; Ming et al. 2020). It has been determined in the literature that cutting performance and wear resistance are 3–10 times higher than tungsten carbide cutters. In particular, silicon nitride ceramic bearings used in the aviation industry have superior speed performance. Ceramic materials are generally evaluated in two groups according to conductive and non-conductive status. While SiC, Al_2O_3, Si_3N_4, ZrO_2 are non-conductive, TiB_2, ZrB_2, B_4C, metal nitrides (TiN and ZrN) are conductive ceramics (Ming et al. 2020). Ceramic cutting tools are ideal for cutting difficult-to-machine materials (Pratap et al. 2019). It exhibits high performance in the machining of cast iron, super alloys, and materials resistant to high cutting temperatures without chemical decomposition (Huang et al. 2020; Grigoriev et al. 2021). Today, the demand for good product quality is increasing while the manufacturing industry is getting pressed between high production and low cost. The metal removal rate (MRR) determined by the cutting speed (CS), feed rate (FR), and depth of cut (DoC) directly affects the machining cost. Cost in machining is related to the interaction between MRR and tool wear (TW). The high FR, CS, and DoC are chosen for high MRR resulting in high heat generation. Although a high MMR ratio is desired in manufacturing, high temperature concentrated in a small area due to the plastic deformation of the material required for chip formation and the friction between the tool, chip, and workpiece during the cutting process affects the workpiece, TW, tool life, workpiece surface integrity, chip morphology, and dimensional stability of the product. In this case, we encounter the MRR and TW, which affect each other in opposite directions. The amount of cutting tool wear requires controlling the friction

DOI: 10.1201/9781003352402-5

between the chip and the tool. This can be achieved by controlling the high temperatures produced in the cutting tool edge region for an efficient machining process. For this reason, many processes cannot be performed efficiently without cooling in the machining process. Cutting fluids are widely used in traditional manufacturing processes as they reduce TW and increase tool life. Cutting fluids cause a cost of 7%–17% of the total cost in the machining process (Klocke and Eisenblätter 1997). In addition, traditional petroleum-based cutting fluids are known to cause many environmental problems due to their chemical decomposition at high temperatures. Therefore, many traditional and non-traditional green machining technologies can be considered alternatives to reduce or completely eliminate cutting fluids. Studies have been carried out by different researchers to reduce or completely eliminate the use of chemical coolants in existing conventional processing methods. In these studies, high-pressure coolants (HPC), cryogenic cooling (CC), solid lubricants, air/gas/steam coolants, and the minimum amount of lubrication or near-dry machining (NDM) methods were presented as alternatives to conventional coolant techniques. Alternative non-traditional machining methods are also used in ultrasonic machining, spark erosion, laser beam machining (LBM), and abrasive water jet machining (AWJM).

In this review study, literature on the machining of ceramic materials with different machinability properties under environmental conditions is presented. The literature presented consists of two main parts i.e. the first is alternative applications on reducing the use of cutting fluid, which is harmful to the environment/human health and causes machinability costs in traditional machining methods, and the second consists of the use of non-traditional machining methods in which the use of cutting fluids is reduced or not used in ceramic structures, which are difficult to cut with traditional methods. Thus, it is aimed to provide an in-depth contribution to the current literature and to provide support for the solution of problems that may be encountered in applications for industrial users.

5.2 GREEN TECHNIQUES FOR CONVENTIONAL MACHINING OF CERAMICS

Coolants are mainly used for lubrication, cooling, and removal of chips in manufacturing. For this purpose, emulsions and standard oils are widely preferred. Intensive use of machining fluid leads to high machining costs. In addition, since it contains a large number of harmful substances, it poses a serious danger to the environment and human health (Roy et al. 2019). Compressed air for cooling is another choice. However, in this case, there is no lubrication feature. Therefore, a more efficient cooling system is needed to ensure dry or near-dry and contamination-free machining in modern manufacturing. A very small amount of cutting fluid (Drop by drop) in the MQL method (Figure 5.1) is sprayed directly into the cutting area with compressed air with the help of a nozzle. In this method, cutting performance can be increased without the use of large volumes of fluid flow (Gupta and Sood 2017). The transport of a small amount of coolant to the cutting zone is accomplished without air or an air medium in MQL. The cutting fluid in the airless system is fed in the form of microdroplets with the help of a

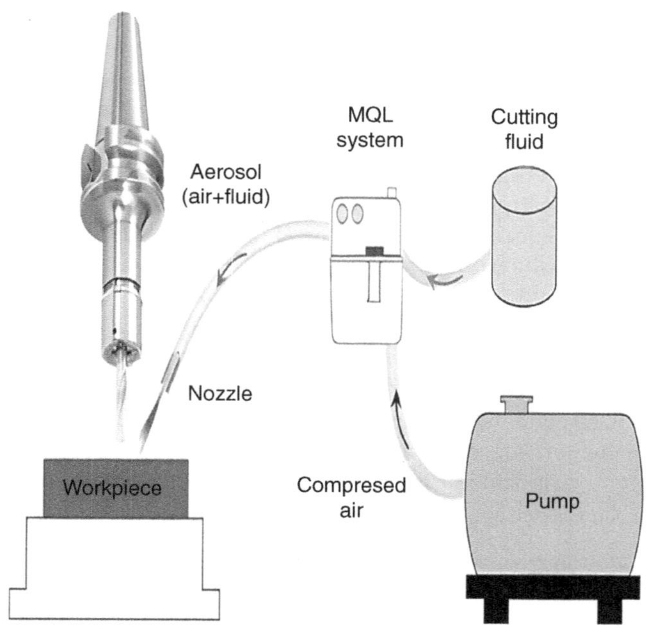

FIGURE 5.1   Schematic representation of MQL

pump. A small amount of cutting fluid mixed with air in the air system is delivered as an aerosol spray from a nozzle (Weinert et al. 2004).

Environmentally friendly MQL technique reduces fluid consumption by up to 0.001 compared to conventional practice (Rabiei et al. 2017). A new MQL method called nanoparticle minimum quantity lubrication (NMQL) is used to increase heat transfer and reduce friction in the cutting zone. In this method, nanoparticles with high thermal conductivity (Hexagonal boron nitride-hBN, graphite, and Al_2O_3) are added to the lubricating fluid (Sharma et al. 2016). The quality of nanofluids plays an important role in the transport of heat along with the improvement in surface quality (Nandakumar et al. 2019). Although it increases the thermophysical properties, the application of nanofluids compromises the cost and short life of the nano-lubricant. While a combination of fluid and nanoparticles is used due to better heat transfer properties, sometimes more than one nanofluid (hybrid) is mixed to improve thermophysical properties (Babu et al. 2017). In a study in the literature, the oblique cutting performance of 0.1–0.6% nanoparticle admixture based on pure coconut oil through MQL of Al-7079/7 wt-TiC composite was investigated. The addition of more than 0.4% nanoparticles reduced MQL performance and significantly increased cutting zone temperature, CFs, TW, and surface roughness (SR) (Sujith and Mulik 2022). MQL reduces the friction between the tool and workpiece in contrast to dry machining conditions. Today, the effects on the machinability properties of ceramic materials with different machinability properties have been investigated. Trueman et al. used different modes called spalling for thermal machining of advanced CMs such as SiAlON-TiN and SiC-TiB$_2$, instead of traditional electrical discharge sinking. They demonstrated that arc-related discharges promote spalling areas caused by thermal shock in the ceramic surface layers. Additionally, by using carbide powder for machining non-conductive CMs, they achieved noticeable results different from the true nature of electrical erosion (Trueman and Huddleston 2000). Banu et al. examined the MRR and recast layer hardness of non-conductive ZrO$_2$ after micro-EDM with a tungsten tool and copper adhesive auxiliary electrode in hydrocarbon liquid. They found that the hardness of the recast layer increased as the gap voltage decreased. Better machinability properties were obtained by immersing the ZrO$_2$ in dielectric fluid using copper adhesive, positive workpiece polarity, feed of 3 μm/s and unidirectional circulation (Banu et al. 2014). Kumar et al. researched the impact of machining parameters on machining outputs in order to cut non-conductive ZrO$_2$ ceramic material with ECDM. They found that DC supply voltage and electrolyte concentration (EC) had a significant impact on machining performance and that 65 V, 16 g/l EC and 60 mm inter electrode gap should be used for better MRR (Kumar et al. 2021). Pu et al. investigated the impact of different operational parameters (CS, FR, DoC, laser power, preheating time, diameter of laser facula, laser-tool distance, and laser-tool angle) on the surface structure in the LAM process of Si$_3$N$_4$ CMs under different MMR modes. It was observed that when the material was cut in plastic, the chips always had a band-like structure. They found that tool wear occurred on the tool rake face and the surface was smooth. While it was observed that the surface became rougher as the laser power decreased, they revealed that as the laser power increased, the wear resistance increased and fluid retainability decreased (Pu et al. 2020). Qu et al. stated the SR and grinding forces of unidirectional carbon fiber-reinforced ceramic matrix (Cf/SiC) material under dry, wet, and MQL conditions. According to the lubrication mechanism of MQL, nozzle direction, air pressure (AP), OFR, and nozzle distance were examined (Figure 5.2). They observed that the temperature of the MQL conditions can be greatly reduced and effective oil films are formed in the contact areas between the material surface and the grit. Compared to other grinding ambient conditions, MQL was found to significantly improve the surface quality and reduce grinding force (Qu et al. 2020a).

In another study, Qu et al. researched the impact of dry, flood, MQL, and nanofluid MQL conditions on machinability characteristics in the grinding process of unidirectional carbon fiber-reinforced ceramic matrix materials. It was determined that excellent cooling and lubrication could be achieved with nanofluid MQL conditions leading to small grinding forces, high surface quality, and low subsurface damage (Qu et al. 2020b). Bayat et al. determined that the temperature was decreased by 16%–35% and the SR by 46% in the grinding process of ZrO$_2$ bioceramics for MQL situations compared to dry situations. However, as a result of the increase in the DoC and the lubricant not penetrating the cutting zone effectively, the cooling and lubrication effects were reduced and MQL performance decreased (Bayat et al. 2022). Kannan et al. examined the impact of friction and heat generation on Al7075/Al$_2$O$_3$ material turning in dry and MQL situations. They stated that CFs and TW were reduced with MQL and improved surface quality by helping chip breakage (Kannan et al. 2020). Davim et al. investigated the machinability properties of A356/20/SiCp-T6 metal matrixed in the turning process using MQL with different

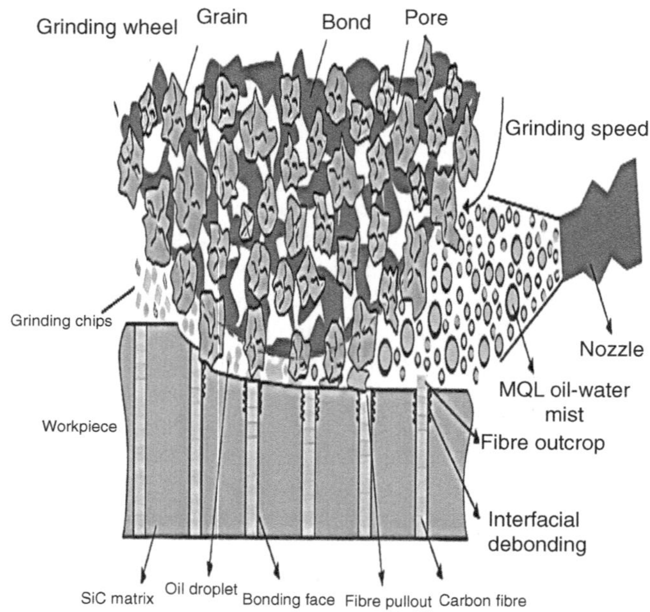

FIGURE 5.2 Schematic representation of grinding Cf/SiC material under MQL conditions (Qu et al. 2020a)

flow rates (between 50 and 2,000 ml/h). They found that different flow rates had a significant impact on the change in CF, while the SR changed little as the flow rate increased (Davim et al. 2009). Duan et al. examined the machinability properties of Al/SiCp materials under different cooling and lubrication conditions (Dry, liquid nitrogen-LN$_2$, MQL, cutting oil, and emulsion) in the turning process. They revealed that SiC particles and chips due to the excellent flushing property of MQL and LN$_2$ were removed from the boundary regions and boundary wear could be reduced. It was observed that MQL and LN$_2$ reduced major flank wear due to their excellent lubricating and cooling properties and tool life in MQL and emulsion usage was longer than in LN$_2$ and oil conditions. In addition, it was stated that hard SiC particles constantly abraded the cutting tool surfaces due to the low lubricating effect of oil and emulsion (Duan et al. 2019). Wang et al. developed a LN$_2$ circulation system to control the cutting tool temperature during turning of Si$_3$N$_4$ ceramic material with PCBN tools. They found that with the LN cooling system, tool wear and surface roughness were importantly decreased compared to normal cutting (Wang et al. 1996). Ma et al. revealed that the material removal method changed from ductile cutting to brittle fracture as the CS or DoC increased during turning of fluorophlogopite CMs using a PCD tool. The critical CS and DoC values for the ductile-brittle transition rate were determined as 47.1 m/min and 0.05 mm, respectively (Ma et al. 2019). Yan et al. observed that PCD tools exhibited superior machinability properties compared to other tools in the machining of alumina-based ceramics containing ZrO$_2$ and TiO$_2$ with PCD, CBN, Al$_2$O$_3$ ceramic and carbide (K10) tools. It was determined that carbide and ceramic tools were not suitable for processing CMs. The optimum cutting parameters were determined as PCD tool CS of 60 m/min, FR of 0.029 mm/rev and DoC of 0.015 mm. They observed that during the machining of CMs in a cool and humid cutting environment, the tool surface became moist and inducted wear on the tool. Additionally, they determined that hot blowing and sucking facilitated chip evacuation and reduced tool wear (Yan et al. 1995). Other green techniques such as dry cutting (Bayraktar 2021), cryogenic cooling-based machining (Bayraktar 2020; Cagan and Buldum 2021), and usage of green lubricants such as vegetable oils, etc., have also been attempted to machine ceramics sustainably.

Thermal assisted machining (TAM) simplifies machining by changing the deformation behavior of the workpiece by increasing the temperature during cutting in an area close to the cutting zone with an external heat source. Laser-assisted machining (LAM) is based on the laser heating of a small area in front of the workpiece. This can be achieved by placing a laser at a single point just in front of the cutting tool in turning and milling operations (Figure 5.3). However, the laser spot size limits the practical application of LAM as it causes little or partial overheating of the workpiece (Xu et al. 2020). Heat-assisted machining contributes to the reduction of CFs. It also improves surface quality by increasing the MRR (Mruthunjaya and Yogesha 2021).

Current studies in the literature on TAM continue. In some of them; Pfefferkorn et al. used a continuous wave CO_2 laser with a wavelength of 10.6 mm, a CS of 0.4 m/s, a FR of 0.02 mm/rev and a laser beam diameter of 3.4 mm in the machining of Si_3N_4 ceramic material produced by isostatic pressing with thermal assisted turning.

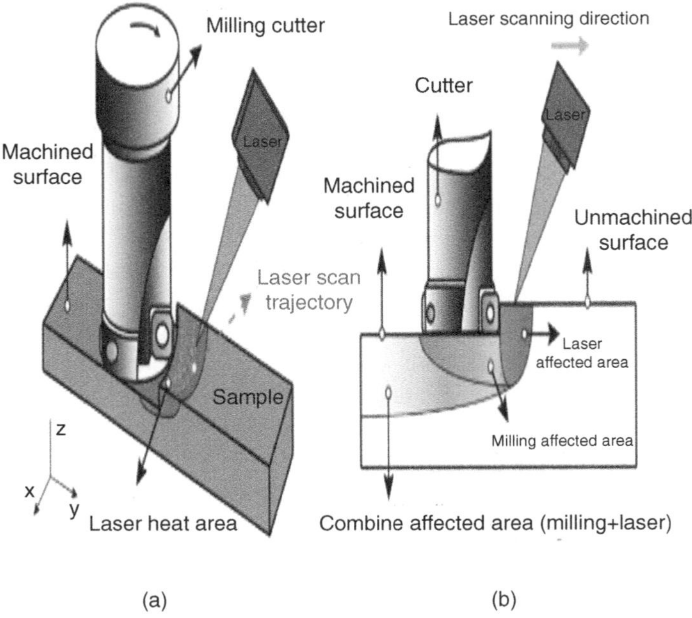

FIGURE 5.3   Heat zones in LAM (Xu et al. 2020)

They demonstrated that a significant reduction in specific energy could be achieved by improving laser heating efficiency (Pfefferkorn et al. 2009). Tian and Shin proposed a thermal modeling approach for LAM in Si_3N_4 ceramic material. For this, they used different laser power (260, 280 and 300 W), spindle speed (500, 1,000 and 1,500 rpm) and feed (0.01; 0.02 and 0.03 mm/rev). They stated that ceramic parts with complex geometry could be produced with the LAM method and that the three-dimensional thermal model obtained provides promising results for surface quality and tool wear (Tian and Shin 2006). Shayan et al. observed that the specific energy and hardness of the material decreased during LAM in SiC ceramics compared to traditional processing methods. They also found that as temperature increases, mechanical energy decreases, and heat release increases (Shayan et al. 2009). Chang and Kuo carried out an experimental study to examine the surface roughness in the LAM-Turning process of Al_2O_3 ceramics with Nd: YAG laser. They found that spindle speed, DoC and CF had a significant impact on LAM performance, and the most important advantage of LAM over conventional machining was higher MRR and better surface quality with moderate tool wear (Chang and Kuo 2007). Konig and Zabolicki applied the LAM process to ceramics in both turning and milling processes. They revealed low CF, small tool wear and high MRR rate as characteristic features of the LAM process (Konig and Zabolicki 1993).

5.3 GREEN TECHNIQUES FOR NON-CONVENTIONAL MACHINING OF CERAMICS

Engineering ceramics have low density, hardness, and wear resistance. EDM is a thermoelectric process that erodes with a series of discrete electrical sparks (Chaubey and Gupta 2023; Chaubey et al. 2023) between the workpiece material and the electrode placed in a dielectric fluid (Sadagopan and Mouliprasanth 2017). The basic requirement in the EDM process is electrical conductivity. High conductivity provides a high MRR (Qudeiri et al. 2020). However, the dielectric properties of ceramics and nanoceramics meet this requirement only under certain conditions. In recent years, this problem has been overcome by adding a composite or nanocomposite conductor into the matrix of non-conductive ceramics as a secondary phase or by using auxiliary electrodes (Volosova et al. 2020). In some studies, in the literature; Schubert et al. emphasized that a conductive carbon layer should be formed when machining insulating ceramics during the EDM process. They found that the maximum hole depths of ATZ (Alümina toughened zirconia, ZrO_2-Al_2O_3) and Si_3N_4-TiN (Silicon nitride hardened with titanium nitride) ceramic composites were 731 and 605 μm, respectively, under constant machining conditions, and larger machining depths were limited due to their high strength (Schubert et al. 2016). Guo et al. proposed a new copper strip and conductive polished auxiliary electrode for high speed-WEDM (HS-WEDM) on insulating zirconia (Figure 5.4). It was observed that the machining speed reduced due to the increase in the percentage of LRS (Low resistance electrical discharge), while it increased with HRS (High resistance electrical discharge) during machining (Guo et al. 2014).

While Yoo et al. researched the micro-EDM properties of SiC ceramic produced by the hot-pressing method (Yoo et al. 2015), Tani et al. used a new auxiliary electrode material to prevent the fracture conditions of Si_3N_4 insulating ceramics (Tani

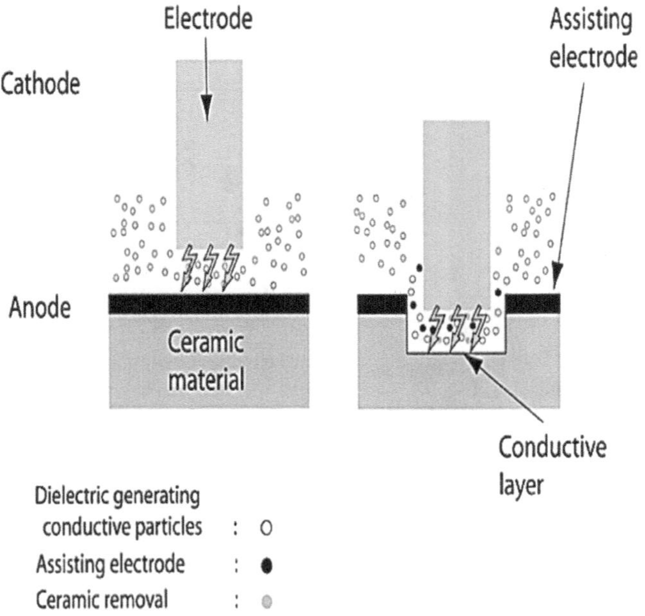

FIGURE 5.4   Spark erosion using auxiliary electrodes (Hösel et al. 2011)

et al. 2004). Zhang studied the effect of WEDM parameters on the surface integrity of conductive TiN/Si_3N_4 ceramics. It was found that TiN/Si_3N_4 ceramics with high TiN content had higher electrical conductivity, which contributed to higher MRR. While the power transistor number increased in direct proportion to the MRR, lower MRR was achieved with higher pulse off-time. Craters, droplets, micropores and microcraks were observed on the machined surfaces. It revealed that the wire electrode material could remain on the machined surface and the combination reaction could occur (Zhang 2014). Solis et al. examined the effect of WEDM parameters (spark gap voltage, pulse-on time, spark frequency, and wire speed) on recast layer thickness (RLT) and surface roughness during the machining of sintered $SiC\text{-}TiB_2\text{-}TiC$ CMs. After statistical analysis, they stated that the most effective parameters on the experimental outputs were pulse-on time and spark gap voltage, respectively. The optimum machining parameters for minimum RLT (3.16 µm) and surface roughness (0.847 µm) were determined as spark gap voltage: 48 V, pulse-on time: 1.0 µs, spark frequency: 10 kHz and wire speed: 8 m/min. As a result of validation tests, they saw a 43.67 and 7.12% reduction in RLT and SR, respectively, compared to initial machining conditions (Solís et al. 2023).

Ultrasonic vibration assisted machining (UVAM) is an unconventional cutting technique that imposes ultrasonic vibrations of very small amplitude and high frequency on the tooltip or workpiece (Figure 5.5).

Three-dimensional contours can be produced simply and quickly with the combination of abrasive liquid (Slurry) by sending ultrasonically induced vibrations to the designed tool. The abrasive liquid must be transmitted to and removed from the

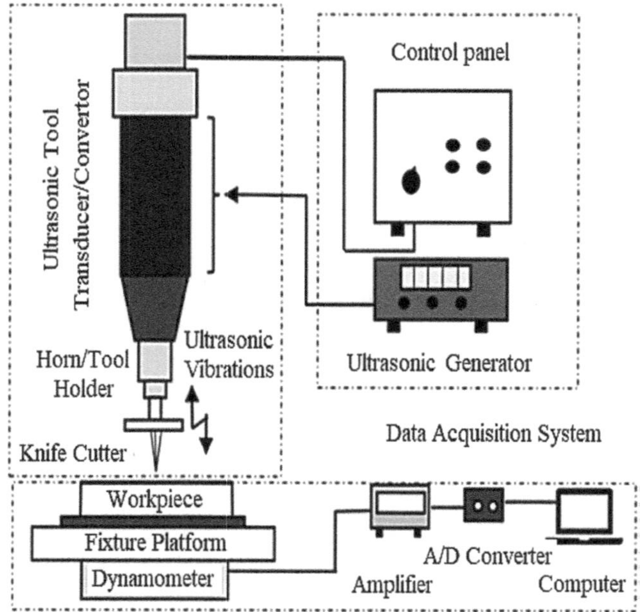

FIGURE 5.5 Schematic of UVAM system (Mughal et al. 2021)

gap between the tool and the workpiece in ultrasonic machining. This process has an impact on MRR and machining performance. In other words, it makes it difficult to drill deep holes and limits dimensional stability in small holes. This problem is eliminated with rotary ultrasonic machining (RUM). However, only drilling is performed with RUM due to the rotational movement of the tool (Pei et al. 1995). The performance of the UVAM system largely depends on the vibration amplitude of the tool obtained by the ultrasonic horn. Higher vibration amplification improves machining quality while minimizing waste (Gupta 2020) and providing a dust-free environment (Mughal et al. 2021). While UVAM is an attractive alternative for materials that are difficult to process with traditional methods, especially ceramic materials, both conductive and non-conductive materials can be machined with this method. Uncut chips are shorter and thinner compared to conventional machining methods. This results in lower CF and higher tool life (Sonia et al. 2021). Therefore, the UVAM method is a widely used method for materials that are difficult to machine, as it improves the dimensional stability of the workpieces (Geng et al. 2020). This process is a proven green and sustainable manufacturing technology with its pollution-free, low power consumption, excellent machining quality and positive contribution to tool life. In some studies in the literature; Zhang et al. developed a new theoretical model for MRR in ultrasonic drilling of engineering ceramics. It was determined that the experimental results obtained were compatible with the model. It was observed that an increase in any of the amplitude of the tool vibration, the applied static load, the size of the abrasive grains, and the rotational speed of the drilling tool caused an increase in MRR. Higher loads often resulted in lower vibration amplitudes due to increased contact time (Zhang et al. 2000). Singh and Singhal examined the impacts of spindle speed, FR, coolant pressure and ultrasonic power parameters on chip size and MRR during RUM of alumina ceramics. As a result of the RSM analysis, they found that the FR on MRR and chip size were the most important parameters and the lowest chip size was obtained with high spindle speed and low FR. They revealed that higher FR, spindle speed and coolant pressure increased MRR (Singh and Singhal 2017). Zeng et al. examined the tool wear in the RUM process of SiC ceramics at constant parameters (Spindle speed: 67 rev/s, FR: 0.09 mm/s, vibration frequency: 20 kHz, coolant pressure: 276 kPa and vibration power supply: 30%). They observed attritional wear and bond fracture formation during RUM. They found that no grain fracture was observed compared to traditional gringing and the wear on the end face was more than the lateral face (Zeng et al. 2005). Li et al. performed finite element analysis and experimental verification for edge-chipping reduction in the RUM process of ceramics. They found that the maximum value of the maximum normal stress and von Mises stress increased as the DoC increased. They revealed that effects of pretightening load did not have a significant effect on the maximum value of maximum normal stress and von Mises stress and that edge-chipping thickness could be decreased by increasing the support length (Li et al. 2006).

Laser beam machining (LBM) is yet another non-conventional or advanced/modern machining technique for green or sustainable machining of ceramics. Materials such as hardened steels and ceramics can be effectively machined with an external heat source that heats and softens the material as it approaches the cutting edge. This is achieved with the aid of a laser or plasma beam directed toward the material surface for cutting (Goindi and Sarkar 2017). LBM is an environmentally friendly and cost-effective machining method as there is no vibration, CF, or TW compared to conventional mechanical machining (Umroh et al. 2020). Cutting quality is the most important factor in laser cutting. Nd: YAG laser or CO_2 lasers are generally used for cutting (Singh and Maurya 2017; Bayraktar and Alparslan 2022a). CO_2 LBM due to features such as good cut quality and high productivity is the ideal non-conventional machining method for machining of ceramic alumina materials (Sharma and Yadava 2018). Current studies on LBM continue in the literature. In some of these, Abdo et al. examined micro channel geometry and surface roughness in the LBM process of zirconia ceramic material. They used laser intensity, pulse frequency, scanning speed, and layer thickness per laser as control parameters. They revealed that high laser intensity (> 90%) and low scanning speed levels (<100 mm/s) caused the laser beam to produce deep V channels and that the layer thickness per laser had a direct effect on the channel geometry. They determined that in order to obtain the ideal geometric size, a layer thickness of 4 μm per laser scan should be used. It was determined that high laser intensity (96%), pulse frequency (15 kHz) and lower scanning speed values (100 mm/s) could be used for minimum surface roughness (Abdo et al. 2019). Kaçar et al. found that alumina ceramics exhibited an approximately linear proportion with crater diameters, pulse duration, and peak power when drilling alumina ceramic using Nd: YAG pulsed laser. In addition, they also revealed that the diameters of the exit holes vary in proportion to the pulse duration and peak power similar to the crater diameters (Kacar et al. 2009). Tian et al. determined that there was a direct proportion between tool wear and machining length in milling silicon nitride ceramics using TiAlN coated tools with the LAM process. They revealed that the cutting forces obtained with LAM were lower than those with conventional machining and the surface quality improved (Tian et al. 2008). Yang et al. determined that the DoC was directly proportional to the laser power in the machining of Al_2O_3 ceramics with the LAM process and that 30 W laser power could be used for better machining quality and deeper cutting depth (Yang et al. 2012). Geethapriyan et al. examined the effects of power (P), transverse speed (TS), standoff distance (SOD), and gas pressure (GP) parameters on MRR, taper (CO), and overcut (OV) outputs in LBM process of alumina ceramics. They observed that the determining parameters during machining were SOD and TS. They found that SOD affected the intensity of the beam to achieve better OV and CO. They determined that TS contributed to higher MRR by reducing cutting time (Geethapriyan et al. 2023).

Abrasive water jet machining (AWJM) can be a prominent solution to the problems and challenges like high cutting fluid consumption, long machining time, and rapid tool wear, in machining of ceramics. AWJM emerges as a good alternative to overcome these problems with a fast and environmentally friendly approach. However, it is known that cutting quality, poor surface quality, kerf taper are the main issues that need to be investigated in difficult to cut materials such as ceramics for AWJM. The cutting surface consists of three zones in the cutting process with AWJM. These are the initial damage region (IDR), smooth cutting region (SCR), and rough cutting region (RCR) (Figure 5.6) (Mayuet et al. 2020; Bayraktar and Alparslan 2022b).

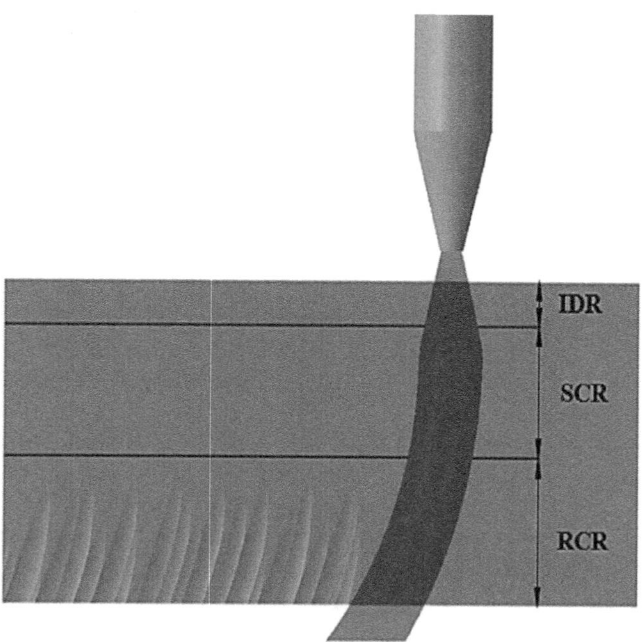

FIGURE 5.6   Schematic representation of cutting zones in cutting with AWJM

The AWJM method is a cold-cutting method. For this reason, it is especially preferred in the machining of workpieces that are not exposed to thermal stresses with an environmentally friendly approach (Bayraktar and Turgut 2018). Looking at recent studies, Ares et al. researched the impact of independent variable parameters (abrasive mass flow rate – AMFR; working pressure – WP; traverse feed rating – TFR) of Al-SiC MMC on taper angle and SR in AWJM. While low feed and high cutting pressures were recommended for minimum taper angle, it was found that the lowest SR (5 µm) could be achieved using low FRs under low and medium WP values (Mayuet et al. 2020). Wang and Guo investigated the cutting performance of industrial alumina ceramics in the AWJM process. They revealed that multipass cutting exhibited significant superiority over single pass cutting with the optimum nozzle TS cutting parameter. They found that surface roughness and kerf taper decreased with the increase in the number of passes (Wang and Guo 2003). Srinivasu et al. examined the effect of operating parameters (FR, impingement angle and number of passes) on the kerf geometry in the AWJM process of silicon carbide ceramics. They revealed that the kerf geometry depends on the change in SOD, abrasive particle velocity, and local impact angles calculated along the footprint. They also found that lower jet feed rates caused the slope of kerf walls to change due to the increase in the depth of erosion (Srinivasu et al. 2009). Wang developed models for predicting depth of jet penetration in the AWJM process of alumina ceramics. He found that nozzle oscillation at small angles with optimum cutting parameters improved DoC by up to 82%. He recommended low oscillation angles (4–6°) and high oscillation frequencies (10–14 Hz) for maximum DoC in the nozzle oscillation cutting process. It was found that the DoC could be predicted with an average error of less than 2.5% under the test situations with the developed model (Wang 2007). Momber et al. studied the machinability of refractory ceramics with magnesia and bauxite abrasives in the AWJM process. They stated that high pump pressure traverse rates and abrasive flow rates were beneficial in terms of cutting performance and that the use of corundum as an abrasive material could increase cutting efficiency compared to garnet (Momber et al. 1996). Wang et al. investigated the effects of jet pressure and TS on the depth of kerf and morphology in the AWJM process of zirconia ceramic tubes. They revealed that depth of kerf increased with increasing jet pressure and decreasing TS and that there was a difference between inlet and outlet kerf morphologies (Wang et al. 2023). Selvan et al. investigated the effect of standoff distance, TS, water pressure, and mass flow rate parameters on the DoC in rough AWJM of alumina ceramics. They developed a model with a maximum deviation of approximately 0.8%–3.9% using the relapse/regression investigation strategy and stated that this model could be used successfully (Selvan et al. 2019).

5.4 CONCLUSIONS

Problems arising in industrial applications have been a driving force for engineers to develop new-generation materials. The demands that cannot be met with traditional engineering materials in the aerospace, optics, biomedicine, and other industries are the main reason for the development of ceramic materials with different properties. Machining technologies are needed to use these materials as final products in mechanical systems. Today, the manufacturing industry uses an environmentally friendly manufacturing process called green machining to minimize environmental damage. The industry has also focused on environmentally friendly manufacturing methods for ceramic structures, which have more restrictive machinability properties than traditional engineering materials. Ceramic structures have limited machinability due to their extraordinary properties. In some cases, effective use of non-traditional machining methods may be necessary. In the literature, it is seen that there are many studies on reducing the use of cutting fluids that are harmful to human and environmental health in traditional and non-traditional manufacturing methods of ceramic structures. It has been observed that MQL and cryogenic machining are frequently compared to dry machining in terms of reducing CF, and TW, and improving workpiece surface quality and dimensional stability in conventional manufacturing practices. However, there are also CMQL studies in which cryogenic and MQL applications are used together. While TW has been seen as a major problem in the dry turning of ceramics, it has been stated that there has been no significant difference in SR in the drilling operation. It has been demonstrated that thermal-assisted machining conditions have positive effects on tool life while reducing machining time and CFs for ceramic materials. It has been observed that environmental studies have been carried out for vegetable-based cutting fluids. It has been determined that spark erosion is particularly focused on the machining of non-conductive ceramic materials. It has been observed that UVAM is preferred in cases where electrical conductivity is not taken into account. MRR, dimensional error, SR, and kerf width have been investigated in LBM as it is environmentally friendly compared to conventional machining. AWJM can be preferred in industrial applications where thermal stresses are not desired and for brittle materials such as ceramics due to its fast and environmentally friendly approach.

REFERENCES

Abdo, B., Ahmed, N., El-Tamimi, A., Anwar, S., Alkhalefah, H., Nasr Emad, A. 2019. "Laser beam machining of zirconia ceramic: an investigation of micro-machining geometry and surface roughness." *Journal of Mechanical Science and Technology* 33(4). Springer: 1817–1831.

Ahlhelm, M., Günther, P., Scheithauer, U., Schwarzer, E., Günther, A., Slawik, T., Moritz, T., Michaelis, A. 2016. "Innovative and novel manufacturing methods of ceramics and metal-ceramic composites for biomedical applications." *Journal of the European Ceramic Society* 36(12). Elsevier: 2883–2888.

Babu, R., Kumar, K., Rao, S. 2017. "State-of-art review on hybrid nanofluids." *Renewable and Sustainable Energy Reviews* 77. Elsevier: 551–565.

Banu, A., Ali, M.Y., Rahman, M.A. 2014. "Micro-electro discharge machining of non-conductive zirconia ceramic: investigation of MRR and recast layer hardness." *The International Journal of Advanced Manufacturing Technology* 75. Springer: 257–267.

Bayat, M., Adibi, H., Barzegar, A., Rezaei, S.M. 2022. "Experimental and numerical investigation of heat generation and surface integrity of ZrO_2 bioceramics in grinding process under MQL condition." *Journal of the Mechanical Behavior of Biomedical Materials* 131. Elsevier: 105226.

Bayraktar, Ş. 2020. "Cryogenic cooling-based sustainable machining." *In High Speed Machining*, edited by Kapil Gupta and Konstantinos Salonitis. Amsterdam: Elsevier Academic Press: 223–241.

Bayraktar, Ş. 2021. "Dry cutting: a sustainable machining technology." *In Sustainable Manufacturing*, edited by Kapil Gupta and Joao Paulo Davim. London: Elsevier: 231–257.

Bayraktar, Ş., Alparslan, C. 2022a. "Introduction to gas and solid state laser techniques in cutting process." *In Advanced Engineering of Materials Through Lasers*, edited by Jeyalakshmi Radhakrishan and Sunil Pathak. Cham: Springer International Publishing: 33–54.

Bayraktar, Ş., Alparslan, C. 2022b. "Sustainable abrasive jet machining." *In Advances in Sustainable Machining and Manufacturing Processes*, edited by Kishor Kumar Gajrani, Arbind Prasad, Ashwani Kumar. Boca Raton: CRC Press: 173–188.

Bayraktar, Ş., Turgut, Y. 2018. "Effects of different cutting methods for electrical steel sheets on performance of induction motors." *Proceedings of the Institution of Mechanical Engineers, Part B: Journal of Engineering Manufacture* 232(7). SAGE: 1287–1294.

Cagan, S.C., Buldum, B.B. 2021. "Cryogenic cooling-based sustainable machining." *In Sustainable Manufacturing*, edited by Kapil Gupta, Konstantinos Salonitis. Amsterdam: Elsevier: 259–285.

Chang, C.W., Kuo, C.P. 2007. "Evaluation of surface roughness in laser-assisted machining of aluminum oxide ceramics with Taguchi method." *International Journal of Machine Tools and Manufacture* 47 (1). Elsevier: 141–147.

Chaubey, S.K., Gupta, K. 2023. "A review on Wire-EDM of bio titanium." *Reports in Mechanical Engineering* 4 (1): 141–152.

Chaubey, S.K. Gupta, K. Madić, M. 2023. "An investigation on mean roughness depth and material erosion speed during manufacturing of stainless-steel miniature ratchet gears by Wire-EDM." *Facta Universitatis, Series: Mechanical Engineering* 21 (2): 239–258.

Chawla, K. 2013. *Ceramic Matrix Composites*. Malta: Springer Science & Business Media.

Davim, J.P., Sreejith, P., Silva, J.V.D. 2009. "Some studies about machining of MMC'S by MQL (Minimum Quantity of Lubricant) conditions." *Advanced Composites Letters* 18(1). SAGE: 21–23.

Duan, C., Sun, W., Che, M., Yin, W. 2019. "Effects of cooling and lubrication conditions on tool wear in turning of Al/SiCp composite." *The International Journal of Advanced Manufacturing Technology* 103(1). Springer: 1467–1479.

Geethapriyan, T., Kailaash Pandiyan, C., Palani, I.A. 2023. "Impact of process parameters on laser beam machining of ceramic material." *In International Manufacturing Science and Engineering Conference* 87240. ASME: V002T06A021.

Geng, D., Liu, Y., Shao, Z., Zhang, M., Jiang, X., Zhang, D. 2020. "Delamination formation and suppression during rotary ultrasonic elliptical machining of CFRP." *Composites Part B: Engineering* 183. Elsevier: 107698.

Goindi, G.S., Sarkar, P. 2017. "Dry machining: a step towards sustainable machining–Challenges and future directions." *Journal of Cleaner Production* 165. Elsevier: 1557–1571.

Grigoriev, S.N., Hamdy, K., Volosova, M.A., Okunkova, A.A., Fedorov, S.V. 2021. "Electrical discharge machining of oxide and nitride ceramics: a review." *Materials & Design* 209. Elsevier: 109965.

Guo, Y., Hou, P., Shao, D., Li, Z., Wang, L., Tang, L. 2014. 'High-speed wire electrical discharge machining of insulating zirconia with a novel assisting electrode." *Materials and Manufacturing Processes* 29(5). Taylor & Francis: 526–531.

Gupta, K. 2020. "A review on green machining techniques." *Procedia Manufacturing* 51. Elsevier: 1730–1736.

Gupta, M., Sood, P.K. 2017. "Surface roughness measurements in NFMQL assisted turning of titanium alloys: an optimization approach." *Friction* 5(2). Springer: 155–170.Hösel, T., Müller, C., Reinecke, H. 2011. "Spark erosive structuring of electrically nonconductive zirconia with an assisting electrode." *CIRP Journal of Manufacturing Science and Technology* 4(4). Elsevier: 357–361.

Huang, Z., Dai, Y., Li, Z., Zhang, G., Chang, C., Ma, J. 2020. "Investigation on surface morphology and crystalline phase deformation of Al80Li5Mg5Zn5Cu5 high-entropy alloy by ultra-precision cutting." *Materials & Design* 186. Elsevier: 108367.

Kacar, E., Mutlu, M., Akman, E., Demir, A., Candan, L., Canel, T., Günay, V., Sınmazcelik, T. 2009. "Characterization of the drilling alumina ceramic using Nd: YAG pulsed laser." *Journal of Materials Processing Technology* 209 (4). Elsevier: 2008–2014.

Kannan, C., Chaitanya, V., Padala, D., Reddy, L., Ramanujam, R., Balan, A.S. 2020. "Machinability studies on aluminium matrix nanocomposite under the influence of MQL." *Materials Today: Proceedings* 22. Elsevier: 1507–1516.

Klocke, F., Eisenblätter, G. 1997. "Dry cutting." *CIRP Annals* 46(2). Elsevier: 519–526.

Konig, W., Zaboklicki, A.K. 1993. "Laser-assisted hot machining of ceramics and composite materials." *Proceedings of the International Conference on Machining of Advanced Materials* 847. NIST Special Publication: 455–463.

Kumar, M., Vaishya, R.O., Suri, N.M., Manna, A. 2021. "An experimental investigation of surface characterization for zirconia ceramic using electrochemical discharge machining process." *Arabian Journal for Science and Engineering* 46. Springer: 2269–2281.

Laouissi, A., Yallese, M.A., Belbah, A., Belhadi, S., Haddad, A. 2019. "Investigation, modeling, and optimization of cutting parameters in turning of gray cast iron using coated and uncoated silicon nitride ceramic tools. Based on ANN, RSM, and GA optimization." *The International Journal of Advanced Manufacturing Technology* 101 (1–4). Springer: 523–548.

Li, Z.C., Cai, L.W., Pei, Z.J., Treadwell, C. 2006. "Edge-chipping reduction in rotary ultrasonic machining of ceramics: finite element analysis and experimental verification." *International Journal of Machine Tools and Manufacture* 46 (12–13). Elsevier: 1469–1477.

Ma, L., Cai, C., Tan, Y., Gong, Y., Zhu, L. 2019. "Theoretical model of transverse and longitudinal surface roughness and study on brittle-ductile transition mechanism for turning Fluorophlogopite ceramic." *International Journal of Mechanical Sciences* 150. Elsevier: 715–726.

Mayuet, A., Pedro, F., Rodriguez-Parada, L., Gomez-Parra, A., Batista, P.M. 2020. "Characterization and defect analysis of machined regions in Al-SiC metal matrix composites using an abrasive water jet machining process." *Applied Sciences* 10 (4). MDPI: 1512.

Ming, W., Jia, H., Zhang, H., Zhang, Z., Liu, K., Du, J., Shen, F., Zhang, G. 2020. A comprehensive review of electric discharge machining of advanced ceramics." *Ceramics International* 46 (14). Elsevier: 21813–21838.

Momber, A.W., Eusch, I., Kovacevic, R. 1996. "Machining refractory ceramics with abrasive water jets." *Journal of Materials Science* 31. Springer: 6485–6493.

Mruthunjaya, M., Yogesha, K.B. 2021. "A review on conventional and thermal assisted machining of titanium based alloy." *Materials Today: Proceedings* 46. Elsevier: 8466–8472.

Mughal, K.H., Qureshi, M.A.M., Raza, S.F. 2021. "Novel ultrasonic horn design for machining advanced brittle composites: a step forward towards green and sustainable manufacturing." *Environmental Technology & Innovation* 23. Elsevier: 101652.

Nandakumar, A., Rajmohan, T., Vijayabhaskar, S. 2019. "Experimental evaluation of the lubrication performance in MQL grinding of nano SiC reinforced al matrix composites." *Silicon* 11 (6). Springer: 2987–2999.

Pei, Z.J., Prabhakar, D., Ferreira, P., Haselkorn, M. 1995. "Rotary ultrasonic drilling and milling of ceramics." *Ceramic Transactions: Ceramic Components* 49. 1–12.

Pfefferkorn, F.E., Lei, S., Jeon, Y., Haddad, G. 2009. "A metric for defining the energy efficiency of thermally assisted machining." *International Journal of Machine Tools and Manufacture* 49 (5). Elsevier: 357–365.

Pratap, A., Patra, K., Dyakonov, A.A. 2019. "A comprehensive review of micro-grinding: emphasis on toolings, performance analysis, modeling techniques, and future research directions." *The International Journal of Advanced Manufacturing Technology* 104. Springer: 63–102.

Pu, Y., Zhao, Y., Zhang, H., Zhao, G., Meng, J., Song, P. 2020. "Study on the three-dimensional topography of the machined surface in laser-assisted machining of Si3N4 ceramics under different material removal modes." *Ceramics International* 46 (5). Elsevier: 5695–5705.

Qu, S., Gong, Y., Yang, Y., Sun, Y., Wen, X., Qi, Y. 2020a. "Investigating minimum quantity lubrication in unidirectional Cf/SiC composite grinding." *Ceramics International* 46 (3). Elsevier: 3582–3591.

Qu, S., Gong, Y., Yang, Y, Wang, W., Liang, C., Han, B. 2020b. "An investigation of carbon nanofluid minimum quantity lubrication for grinding unidirectional carbon fibre-reinforced ceramic matrix composites." *Journal of Cleaner Production* 249. Elsevier: 119353.

Qudeiri, J.E.A., Zaiout, A., Mourad, A.H., Abidi, M.H., Elkaseer, A. 2020. "Principles and characteristics of different EDM processes in machining tool and die steels." *Applied Sciences* 10 (6). MDPI: 2082.

Rabiei, F., Rahimi, A.R., Hadad, M.J. 2017. "Performance improvement of eco-friendly MQL technique by using hybrid nanofluid and ultrasonic-assisted grinding." *The International Journal of Advanced Manufacturing Technology* 93 (1). Springer: 1001–1015.

Roy, S., Kumar, R., Sahoo, A.K., Das, R.K. 2019. "A brief review on effects of conventional and nano particle based machining fluid on machining performance of minimum quantity lubrication machining." *Materials Today: Proceedings* 18. Elsevier: 5421–5431.

Sadagopan, P., Mouliprasanth, B. 2017. "Investigation on the influence of different types of dielectrics in electrical discharge machining." *The International Journal of Advanced Manufacturing Technology* 92 (1–4). Springer: 277–291.

Schubert, A., Zeidler, H., Kühn, R., Hackert-Oschätzchen, M., Flemmig, S., Treffkorn, N. 2016. "Investigation of ablation behaviour in micro-EDM of nonconductive ceramic composites ATZ and Si3N4-TiN." *Procedia CIRP* 42. Elsevier: 727–732.

Selvan, C.P., Madara, S.R., Pillai, S.R., Vandanapu, R. 2019. "In-depth evaluation of process variables in abrasive waterjet cutting of alumina ceramics." *In 2019 Advances in Science and Engineering Technology International Conferences (ASET)*. IEEE: 1–4.

Sharma, A.K., Tiwari, A.K., Dixit, A.R. 2016. "Effects of Minimum Quantity Lubrication (MQL) in machining processes using conventional and nanofluid based cutting fluids: a comprehensive review." *Journal of Cleaner Production* 127. Elsevier: 1–18.

Sharma, A., Yadava, V. 2018. "Experimental analysis of Nd-YAG laser cutting of sheet materials–A review." *Optics & Laser Technology* 98. Elsevier: 264–280.

Shayan, A.R., Poyraz, H.B., Ravindra, D., Patten, J.A., Ghantasala, M. 2009. "Force analysis, mechanical energy and laser heating evaluation of scratch test on silicon carbide in micro-laser assisted machining process." *International Manufacturing Science and Engineering Conference* ASME: 1–6.

Singh, R.P., Singhal, S. 2017. "Investigation of machining characteristics in rotary ultrasonic machining of alumina ceramic." *Materials and Manufacturing Processes* 32 (3). Taylor & Francis: 309–326.

Singh, S.K., Maurya, A.K. 2017. "Review on laser beam machining process parameter optimization." *International Journal for Innovative Research in Science & Technology* 3 (8). 34–38.

Solís, P.N.W., Malakhinsky, A., Soe, T.N., Pristinskiy, Y., Smirnov, A., Meleshkin, Y., Apelfeld, A., Peretyagin, N., Peretyagin, P., Grigoriev, S.N. 2023. "Investigation of the WEDM parameters' influence on the recast layer thickness of spark plasma sintered SiC-TiB2-TiC ceramic." *Coatings* 13 (10). MDPI: 1728.

Sonia, P., Jain, J.K., Saxena, K.K. 2021. "Influence of ultrasonic vibration assistance in manufacturing processes: a Review." *Materials and Manufacturing Processes* 36 (13). Taylor & Francis: 1451–1475.

Srinivasu, D.S., Axinte, D.A., Shipway, P.H., Folkes, J. 2009. "Influence of kinematic operating parameters on kerf geometry in abrasive waterjet machining of silicon carbide ceramics." *International Journal of Machine Tools and Manufacture* 49 (14). Elsevier: 1077–1088.

Sujith, S.V., Mulik, R.S. 2022. "Surface Integrity and Flank Wear Response Under Pure Coconut Oil-Al_2O_3 Nano Minimum Quantity Lubrication Turning of Al-7079/7 wt%-TiC In Situ Metal Matrix Composites." *Journal of Tribology* 144 (5). ASME: 051701.

Tani, T., Fukuzawa, Y., Mohri, N., Saito, N., Okada, M. 2004. "Machining phenomena in WEDM of insulating ceramics." *Journal of Materials Processing Technology* 149 (1–3). Elsevier: 124–128.

Tian, Y., Shin, Y.C. 2006. "Thermal modeling for laser-assisted machining of silicon nitride ceramics with complex features." *Journal of Manufacturing Science and Engineering* 128. ASME: 425–434.

Tian, Y., Wu, B., Anderson, M., Shin, Y.C. 2008. "Laser-assisted milling of silicon nitride ceramics and Inconel 718." *Journal of Manufacturing Science and Engineering* 130 (3). ASME: 031013.

Trueman, C.S., Huddleston, J. 2000. "Material removal by spalling during EDM of ceramics." *Journal of the European Ceramic Society* 20 (10). Elsevier: 1629–1635.

Umroh, B., Ginting, A.İ, Rahman, M.N.A. 2020. "CO_2 laser machining on alumina ceramic: a review." *In IOP Conference Series: Materials Science and Engineering* 1003 (1). IOP Publishing: 012131.

Volosova, M., Okunkova, A., Fedorov, S., Hamdy, K., Mikhailova, M. 2020. "Electrical discharge machining non-conductive ceramics: combination of materials." *Technologies* 8 (2). MDPI: 32.

Wang, J. 2007. "Predictive depth of jet penetration models for abrasive waterjet cutting of alumina ceramics." *International Journal of Mechanical Sciences* 49 (3). Elsevier: 306–316.

Wang, J., Guo, D.M. 2003. "The cutting performance in multipass abrasive waterjet machining of industrial ceramics." *Journal of Materials Processing Technology* 133 (3). Elsevier: 371–377.

Wang, P., Miao, X., Wu, M., Zhou, P. 2023. "Study on the process of abrasive water jet cutting for zirconia ceramic tubes." *The International Journal of Advanced Manufacturing Technology* 126. Springer: 5555–5569.

Wang, Z.Y., Rajurkar, K.P., Murugappan, M. 1996. "Cryogenic PCBN turning of ceramic (Si3N4). *Wear* 195 (1–2). Elsevier: 1–6.

Weinert, K., Inasaki, I., Sutherland, J., Wakabayashi, T. 2004. "Dry machining and minimum quantity lubrication." *CIRP Annals* 53 (2). Elsevier: 511–537.

Xu, D., Liao, Z., Axinte, D., Sarasua, J.A., M'Saoubi, R., Wretland, A. 2020. "Investigation of surface integrity in laser-assisted machining of nickel based superalloy." *Materials & Design* 194. Elsevier: 108851.

Yan, B.H., Huang, F.Y., Chow, H.M. 1995." Study on the turning characteristics of alumina-based ceramics." *Journal of Materials Processing Technology* 54 (1–4). Elsevier: 341–347.

Yang, J., Yu, J., Cui, Y., Huang, Y. 2012. "New laser machining technology of Al_2O_3 ceramic with complex shape." *Ceramics International* 38 (5). Elsevier: 3643–3648.

Yoo, H.K., Ko, J.H., Lim, K.Y., Kwon, W.T., Kim, Y.W. 2015. "Micro-electrical discharge machining characteristics of newly developed conductive SiC ceramic." *Ceramics International* 41 (3). Elsevier: 3490–3496.

Zeng, W.M., Li, Z.C., Pei, Z.J., Treadwell, C. 2005. "Experimental observation of tool wear in rotary ultrasonic machining of advanced ceramics." *International Journal of Machine Tools and Manufacture* 45 (12–13). Elsevier: 1468–1473.

Zhang, C. 2014. "Effect of wire electrical discharge machining (WEDM) parameters on surface integrity of nanocomposite ceramics." *Ceramics International* 40 (7). Elsevier: 9657–9662.

Zhang, Q.H., Wu, C.L., Sun, J.L., Jia, Z.X. 2000. "The mechanism of material removal in ultrasonic drilling of engineering ceramics." *Proceedings of the Institution of Mechanical Engineers, Part B: Journal of Engineering Manufacture* 214 (9). SAGE: 805–810.

6 Life Cycle Engineering and Analysis of Machining Processes

Suleyman Cinar Cagan

6.1 INTRODUCTION

Today, the manufacturing industry has determined the speed and accuracy of intelligent production and production with industry 4.0 as the primary target. However, it sets the social, environmental, and economic benefits to the second. Industry 5.0 (Manufacturing in artificial intelligence process) is an industrial development aimed at achieving many social goals, including sustainability, using industry 4.0 technologies. In addition to industry 5.0, environmental and human-compatible manufacturing techniques are also one of the latest missions of the industry, "sustainable manufacturing." The technological components of the manufacturing industry have been the subject of numerous research up to this point, and the qualification technically requested is assumed to be achieved. The development of technical and economic issues and other related e such as the environment and human health was ignored. Because the development process requires integrity, essential factors such as the environment and human health must be harmonized in manufacturing to achieve better results. After the great industrial revolution, traditional machining methods reached maturity in the late 1900 years. With the advancement of modern production methods, high-added jobs have become possible. High technology is required for the machining of high-quality products, and this process needs to be economically assessed. Sustainable manufacturing, high-speed machining, and high-accuracy modern manufacturing techniques are the main issues that need to be developed. Specialization in these matters is critical to the manufacturing industry. These factors must have attained the levels required by technology during the great industrial revolution.

In recent years, intensive efforts have been undertaken to use limited energy resources more efficiently and to prevent increased environmental pollution; to reduce the increasing costs over time, such as production, labor, and transportation (Faruk et al., 2017; Santos et al., 2017). Among these studies, work on materials that reduce energy consumption and environmental pollution remains important (Nee, 2014).

The machining of new materials, produced by the continuous development of material technologies, has become a science rather than an industrial practice and is

DOI: 10.1201/9781003352402-6

developed by continuous research (Murray, 1997). The production of new material is a research issue, and the results of this research lead to industrial applications. The manufacturing of a product goes through the following stages: First, the product is designed, then the process is planned by determining which method it can be processed, and finally, the quality control operations are carried out (Garetti and Taisch, 2012). The turning process is one of the most basic and essential methods of traditional material forming (Youssef and El-Hofy, 2022). By turning, the chip is obtained in the desired size and precision by removing materials from cylindrical parts (Davim, 2008). When executing this operation, efficiency is the most important characteristic, and it is preferable to produce the best surface quality of the part at the lowest cost (Davim, 2008). In this case, the idea of sustainable green manufacturing is the target.

Machining can be considered the essential product acquisition of the manufacturing industry (El-Hofy, 2018). The use of life cycle analysis in the manufacturing industry results in the concept of environmentally responsible and sustainable production (Pusavec et al., 2010). This leads in a strategy that is sustainable to design and engineering activities in product and system operations, minimizing environmental impact. This strategy seeks to reduce environmental effect across the whole product's life cycle, including design, production, usage, and disposal, while maximizing throughput from sources (Silva et al., 2015).

6.2 LIFE CYCLE ENGINEERING (LCE)

Life cycle engineering aims to promote economic progress and to protect the environment and resources first. For this purpose, it covers engineering activities that involve applying products to design and manufacture under the leadership of technological and scientific principles (Alting et al., 2007). The main topic that engineers and scientists are focusing on is optimizing the product life cycle from the cradle to the grave, minimizing environmental pollution, and ensuring efficient production (Figure 6.1) (Hauschild et al., 2018; PennState, 2020). LCE is an analysis system that holistically evaluates all life cycles of a product or service and their connections to each other. Environmental impacts are assessed by natural resource consumption, such as climate change, thinning in the stratospheric ozone layer, otrification, acidification, and toxic emissions (Herrmann et al., 2014).

The standard ISO 14040 states that the product life cycle covers all stages between the disposal of raw materials (Finkbeiner et al., 2006). This coverage is double-sided. The output and production phase of raw materials is defined as upward-facing, and the destruction of the product is defined as a downstream process. In today's high consumption, the LCE reduces the negative consequences of consumption to the environment (atmospheric emissions, water-borne pollutants, solid waste, etc.) (Jamwal et al., 2021). It aims to develop products and provide more efficient and sustainable engineering processes to ensure that future generations maintain a quality of life. Efficiency is a classic engineering term that aims to maximize output with minimal input and cost or to get the most out of resources (Madanchi, 2021; Sangwan and Mittal, 2015).

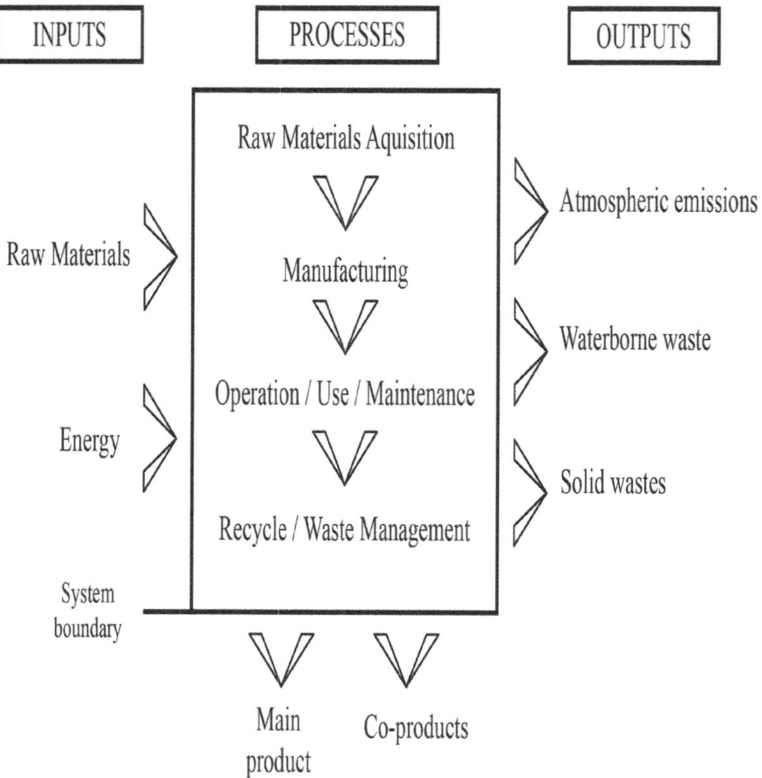

FIGURE 6.1   The life cycle assessment considers the main stages and regular inflows and outflows

6.2.1 Definition, Historical Development, and General Principles

Life cycle analysis is an analysis of the environmental impact at every stage of the process, from the production of a product or process to its disposal, to be analyzed, analyzed, reported, calculated, and managed (Curran, 2012). It is a term that was first introduced in the United States in the years 1950–1960 (Guinee et al., 2011). The World Energy Congress in 1963 also relied on limited raw materials resources and calculations for solving energy resource problems (Kaswan et al., 2020). This means that it has been used to interpret data for energy consumption and resource acquisition in the future as the constraints on energy and raw materials use increase. Later, it was developed as a waste detection system and is located in the environmental management system, managed by the International Standard Organization (ISO), as the ISO 14040 series. In the 1970s, the LCA technique was created in the United States and supported studies and projects designed by the Environmental and chemical toxicology Association and the United States Environmental Protection Agency to create a standard for life cycle inventory analysis and practical evaluation (Bojarska et al., 2021). The standards published by ISO are guided when performing life cycle analysis (Arvanitoyannis, 2008).

6.2.2 Stages and Types of LCE

The life cycle assessment evaluates the environmental effect of a product, process, or service during its manufacturing, usage, and disposal (Shaked et al., 2015). This means that a product's environmental aspects are technically applied to evaluate it throughout its life cycle. The life cycle analysis framework is available in Figure 6.2 (Rebitzer et al., 2004).

An LCA study consists of four stages (Hauschild et al., 2020; Penciuc et al., 2016):

Step 1. LCA goal and scope definition: A product, service, or system is modeled in life cycle analysis. It is crucial to ensure that simplification and distortions do not affect the results of the most fundamental challenge in evaluating the model as a simplified reality. The easiest approach to do this is to explicitly define the goal and scope of LCA investigations. It is the stage in which it is determined what is included and what is not included in the study.

Step 2. Inventory analysis of extractions and emissions: In this step, inventory analysis defines material and energy flow within the production system and its interaction with the environment, particularly raw materials and environmental emissions.

Step 3. Life cycle impact assessment (LCIA): Evaluation of data obtained from inventory analysis is carried out in impact analysis. The display results of all impact categories are detailed in this step. In order to:

- Selecting and defining impact categories – defining relevant environmental impact categories
- Classification – Association of LCA results with impact categories
- Characterization – use the scientific characterization factors to determine each category of impact
- Modeling
- Normalization: describe potential effects that can be compared
- Grouping – classifying and sorting indicators
- Weighting – highlighting the most critical potential effects
- Evaluate and report life cycle impact analysis results

Step 4. Interpretation: The interpretation of a life cycle includes critical review, data sensitivity determination, and results in presentation. This is the process of evaluating energy and material inputs at each stage and opportunities to reduce environmental impact.

According to the relevant ISO standard, the LCA results are interpreted in three stages outlined below (Klöpffer, 2012):

- Definition of essential and critical results in connection with the stages of the LCA

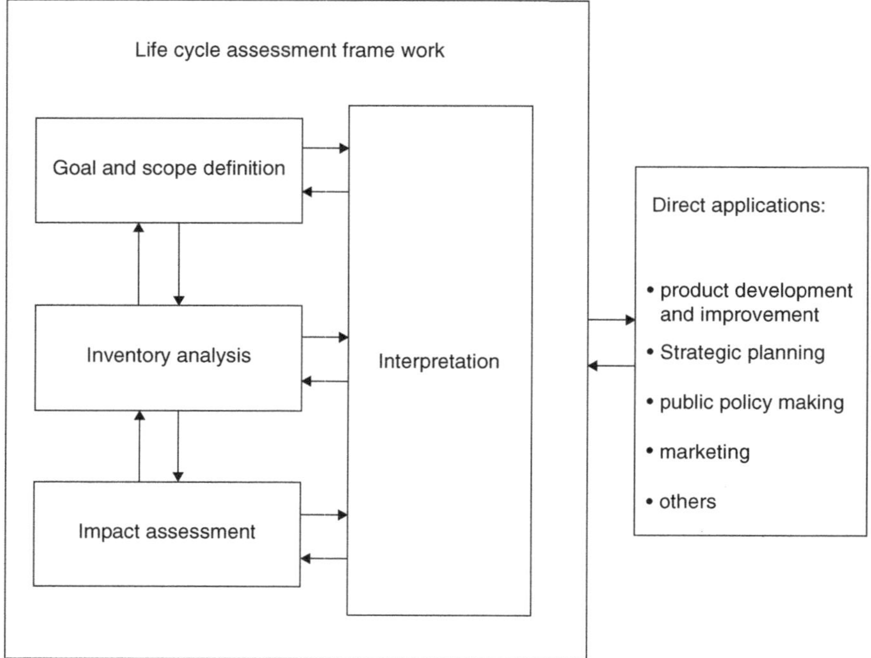

FIGURE 6.2   Life cycle analysis framework (Rebitzer et al., 2004)

- Demonstrate that the work is performed in full, the results are consistent with the initial hypotheses, and that precision analysis is performed at every stage that is deemed necessary, with an overall assessment
- Explain the main results and recommendations in a clear language

When conducting a life cycle assessment study, the following topics should be covered (Sonnemann and Margni, 2015):

The burden of human activity on the environment can be determined by considering resources and energy (Inputs) consumed at every stage of a product's life cycle, pollutants, and waste (Outputs) that arise. The detrimental consequences of renewable and non-renewable resources, human health, and biodiversity on long-term sustainability are then assessed. Once this information is obtained, steps may be made to lessen the environmental effect of printing (or inventories). Figure 6.3 depicts the steps involved in a company's life cycle analysis as a schematic (Guinee et al., 2011).

The results of an LCA can help businesses, policymakers, and other organizations make more informed decisions to move toward sustainability. Provides critical data that can support (Sonnemann and Margni, 2015):

- Improve process and product design
- Marketing (to create environmental awareness and meet consumer demand for green products)

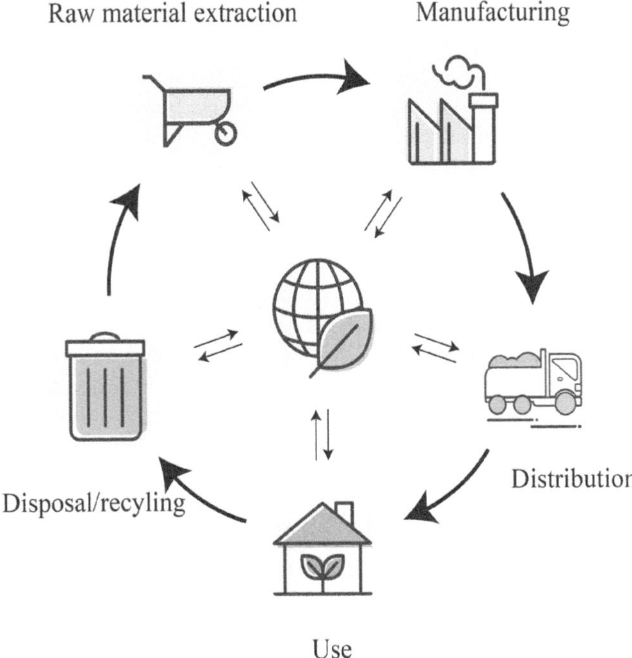

FIGURE 6.3 Processes in life cycle analysis

- Edit continuous improvement
- Certification
- Method for measuring fundamental environmental impacts (greenhouse gas, carbon emissions, water and energy consumption, etc.)
- Target climate change and other sustainability policies

6.2.3 SUSTAINABILITY AND LCE

The term "sustainability" was originally used in the Bruntland Report in 1987, and it continues to be utilized today (Borowy, 2013; Kuhlman and Farrington, 2010). According to the United Nations report 1987, "developments that meet today's needs without compromising the ability to meet the needs of future generations' own needs" (Redclift, 2005).

Although the concept of sustainability is a new idea, one of the most significant topics in the manufacturing business has been the concept of sustainable production in recent decades (Machado et al., 2020; Pimenov et al., 2022). Sustainability has three key components called social, economic, and ecological (Purvis et al., 2019). The ecological column is usually the most obvious. Companies operating on a worldwide scale have increasingly prioritized lowering their carbon footprint, packaging waste, water use, and overall environmental effect (Kravchenko et al., 2019). The key reason is that businesses have discovered that having a positive influence on the environment may also have a favorable financial impact. Another key aspect of sustainability is the social pillar. A sustainable business requires the acceptance and support of its workers, stakeholders, and the community in which it works. The techniques to delivering and sustaining this support vary, but examples include treating employees fairly and giving decent benefits both locally and worldwide. Most firms believe that the economic leg of sustainability is a solid surface (Malek and Desai, 2020). A firm must be lucrative in order to be sustained. Profit, however, cannot outshine the other two columns. Profit at all costs is not the subject of economic foundation. The activities that meet the economic foundation are compliance, appropriate management, and risk management. Sustainability will be achieved when these columns are compatible (Khoshnevis et al., 2017).

Sustainability, smart cities, and technology are provided as a result of a sustainable industry (Energy efficiency, pollution, resources), an environmentally responsible production, a strategic combination of citizen welfare (public safety, education, health, social care), and economic development (investment, business, innovation) (Pacheco Rocha et al., 2022). Smart city and sustainable development policies and objectives must be based on intelligent industries (especially the manufacturing industry) and services (intelligent energy, water, transportation, and buildings) and supported by intelligent infrastructure (sensor networks, data analytics, smart devices, control systems, communications platforms).

Each company must assess its product life cycle and determine whether it is producing sustainable manufacturing. Although not every company can analyze the sub-frame to the grave, they are still evaluating their production. By doing so, they discover problems in production cycles but can determine that their products are energy consumption and are harmful to the environment. This allows them to identify their missing side and try to improve them. However, in recent years, the products of sustainable and environmentally responsible companies have attracted more investors and customers. The sustainability policies of companies are also critical to their development of companies. To reduce environmental impact, it is critical to evaluate product life cycle analysis and create recycling solutions. The more the recycling options available for products, the fewer the products to waste storage, which increases companies' sustainability.

6.3 MACHINING AND LCE

6.3.1 Effect of Machining Parameters on LCE

6.3.1.1 Machining Process Parameters

Machining technology covers a wide range of topics that must be understood to understand and select a specific machining technology accurately. The nature of the machining process employed for a certain material is defined by the cutting tool (Jain, 2009). Parameters that affect operability can generally be specified as cutting speed, feed rate, depth of cut, cutting tool, and environments (Arrazola et al., 2013). As seen on the right side of Figure 6.4, the primary purpose of the adopted technology is to use the machining parameters selected to produce the component at economic and high production rates (El-Hofy, 2018; Youssef and El-Hofy, 2020). Parts must be machined with accuracy, surface texture, and surface integrity, which

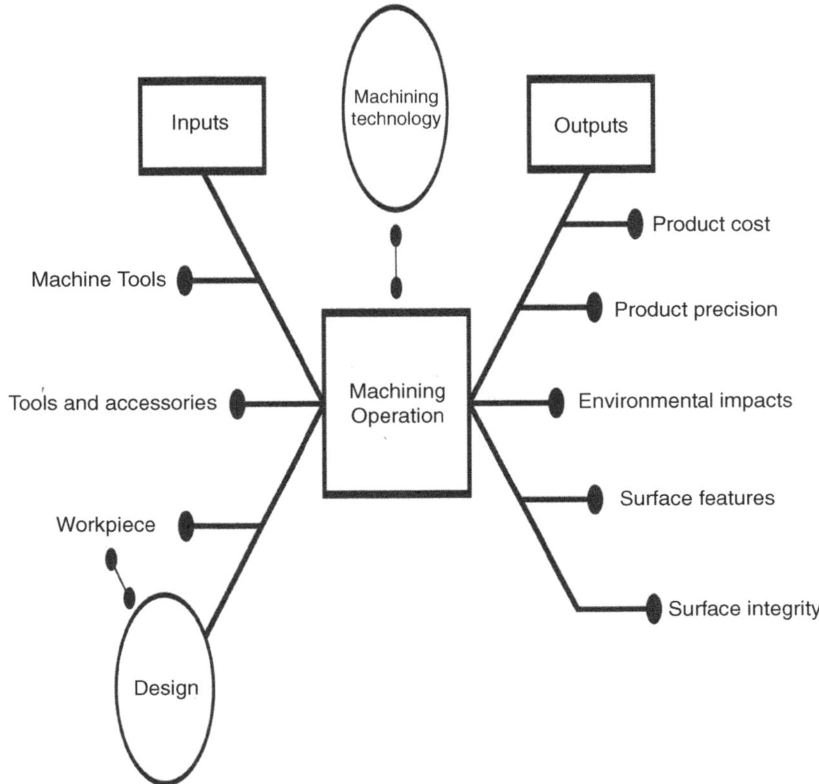

FIGURE 6.4   General aspects of machining technology

satisfies the product designer and eliminate the need for post-machining, ultimately protecting acceptable machining costs (El-Hofy, 2018).

6.3.1.1.1   Cutting Speed

During the turning process, the workpiece rotates at a certain speed in revolutions per minute (Rpm). At the point at which this rotation speed comes into contact with the workpiece of the cutting tool, the circumferential speed, which is generated based on the diameter being processed, is called the cutting speed (V). The cutting speed is the path the cutting tool receives in meters in one minute around the workpiece. In other words, the linear length of the chip that the cutting tool lifts over the workpiece in one minute. The cutting speed (V) can be calculated using Equation (6.1) (Denis, 2005).

$$V = \pi.D.n / 1000 \tag{6.1}$$

Where V is the cutting speed (m/min), D is the diameter (mm) of the workpiece at the point where the cutting tool touches the workpiece, and n refers to the number of revolutions per minute (Rpm) of the workpiece.

The most important thing to consider while choosing the optimal machining conditions for the sample is the cutting speed. The workpiece's material and the cutting tool, as well as the power and condition of the table, and the chip section to be removed are factors that affect the coolant cut-off speed used. Cutting speed is the main factor in increasing and decreasing production capacity. A reduced number of parts produced in unit time occurs at low cutting speeds. In addition, at low cutting speeds, there is an on-team sipping called the micro-welding of the workpiece on the cutting tool, resulting in reduced machining performance. In addition, if the cutting speed is selected too high, it will trigger the formation of high temperatures in the machining area, causing rapid wear of the cutting tool and reduced tool life. As a result, the optimal chip removal cutting speed must be chosen to account for the cutting tool's life and the chip removal ratio (Denis, 2005; Grote and Antonsson, 2009).

6.3.1.1.2 Feed Rate

The feed rate, which refers to the distance traveled by the cutting tool along its axis for each revolution of the workpiece, is a crucial parameter that determines the machined surface quality of the workpiece and the production of the desired chip shape. It is generally affected by the same factors that impact cutting speed, as mentioned previously. The feed rate's magnitude has a significant impact on both the thickness of the chip produced and the breaking of the chip. Higher feed rates result in thinner chips, while lower feed rates result in thicker chips. However, excessively high feed rates can lead to tool wear, while excessively low feed rates can lead to built-up edge formation and poor surface finish. Therefore, the selection of an appropriate feed rate is crucial for achieving optimal machining performance. Furthermore, the feed rate is often dependent on the workpiece material, tool geometry, and cutting conditions, and optimizing it requires careful consideration of these factors (Mills, 2012).

6.3.1.1.3 Depth of Cut

Turning is a key manufacturing process that involves the removal of material from the workpiece to produce a desired shape or surface. The depth of cut, which is defined as the distance between the tool tip and the workpiece surface, plays a critical role in determining the quality of the machined surface and the rate of material removal. The selection of an appropriate depth of cut is influenced by several factors, such as the properties of the workpiece material, the cooling technique used, the type of cutting tool, the rigidity of the system, and the power of the lathe. Therefore, the determination of the depth of cut requires careful consideration of these factors to ensure optimal performance and quality of the final product. Additionally, it is worth noting that excessive depth of cut can result in increased tool wear and vibration, leading to poor surface finish and reduced tool life. Therefore, a balance between the depth of cut and other machining parameters must be maintained to achieve optimal results (Groover, 2020).

6.3.1.1.4 Cutting Tool

The physical chip removal event is removing unwanted material by touching the under-hardened workpiece of the cutting tool (Smith, 2008). This removed layer is called a chip. When removing the chip, the material first has an elastic shape change, then plastic deformation, and finally a break (Shokrani and Biermann, 2020). During this time, the high-pressure build-up of the cutting tool assembly and, consequently, high temperature. Therefore, the unique features of the cutting tool materials are listed in the following items. The stiffness value and pressure strength must be high to allow the cutting tool to penetrate the workpiece and withstand wear. It must have high buckles and high bending strength against pressure and bend during machining. The chip must be resistant to high temperatures from the temperature generated during lifting and should not be likely to soften with temperature (Groover, 2020). The most technically essential features of the cutting tool are hardness and buckshot. Usually, stiffness and toxicity are opposite concepts. Therefore, hard material is generally less toxic. Therefore, research in cutting tool materials is carried out to obtain high-strength raw materials (Panda et al., 2018).

6.3.1.2 Machining Environments

The machining environments are the machining environments required to achieve a part at the desired size and quality. These media are determined by the material type, the cutting material, and basic elements such as machining settings. Dry, wet, MQL, NanoMQL, and cryogenics use many methods in the manufacturing industry. Cutting fluid is used for those other than the dry and cryogenic methods. The cutting fluid's principal role is to provide cooling and lubrication between the tool and the workpiece. A metal machining fluid can help to minimize tool wear, increase machine surface integrity, and remove chips from the cutting zone, all of which contribute to long-term metal machining (Gupta et al., 2020a, 2020b). Since the beginning of the manufacturing industry, these cutting fluids have been employed to cut metal (For the last 200 years). Taylor states that in the first quarter of 1900, it applied a large quantity of water was cooler and that the cutting speed in high-speed steel tools and machining steel could be increased by up to 40% (Taylor, 1906). Despite its superb cooling properties, water is a poor lubricant and causes major corrosion difficulties. As a consequence, new coolers with enhanced lubricating characteristics are being commercially developed to lower the cutting force and power required in the machining process while also increasing the item's surface integrity. Depending on the surface coating, the cutting fluid's properties may require more cooling, lubrication, or both. The cutting fluid's efficiency is determined by various aspects, including the kind of machining processes, cutting parameters, and cutting fluid application methods (Madhukar et al., 2016).

The heat generated on the chip interface removed from the material by the cutting tool is removed from the environment in three ways (Benedicto et al., 2017). Some of this work goes into the workpiece, part of it goes into the cutting tool, and part goes into the chips. If the workpiece becomes too hot, it will cause distortions in part due to the part's expansion, and the product's desired quality will not be achieved. The excess heating of the part surface will also alter the mechanical properties of the part. After the cutting tool has overheated, the cutting tool tip may suddenly become damaged, or the tool life may be reduced. As a result, selecting a cutting tool based on the cutting parameters to be used is critical. Heat generation during cutting is a significant concern that needs to be addressed by efficiently removing it from the cutting area along with the chips. The majority of the heat generated during an ideal cutting process, ranging from 74% to 96%, is taken away from the environment by the chips, whereas the cutting tool and workpiece account for 2%–18% and 1%–20% of heat removal, respectively. Hence, the removal of heat from the cutting zone is primarily dependent on the chips, and their effective removal becomes crucial in mitigating the heat generated during the cutting process (Fleischer et al., 2007).

Coolant is the most efficient way to remove heat from the environment. This facilitates the removal of heat from the environment between the cutting tool and the chip. Using suitable coolant/lubrication fluid allows more than 50% of the heat generated to be discarded with the chip. Using suitable cooling/lubrication systems makes cutting operations easier at high cutting speeds and allows increased chip removal volume and production rate.

Figure 6.5 depicts the primary and secondary deformation locations where heat is created during the upright cutting process, and the QC shows the heat discarded with the chip, the heat from the Qt cutting tool, and the temperature delivered to the QW workpiece (Hao and Liu, 2020). The impact of the third deformation zone on heat generation is often disregarded. The increase in cutting temperature within the tool-chip contact area is dependent on various factors, such as the material and physical properties of the workpiece, cutting parameters, and cutting tool geometry. The secondary heat source plays a major role in elevating the temperature at the tool-chip interface, while the primary heat source indirectly contributes to the rise in cutting temperature. A portion of the cutting heat generated in the primary deformation zone is transferred to the chip during the cutting process and subsequently delivered to the

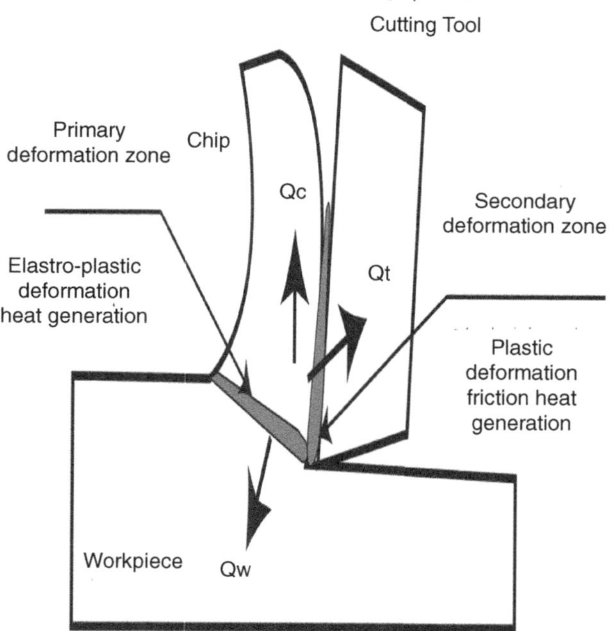

FIGURE 6.5   Heat dissipation during the cutting process

tool-chip surface via the contact interface. The heat partitioning factor determines the amount of cutting heat transmitted to the tool-chip surface through the contact interface (Lawal, 2013).

When the coolant/lubrication fluid is not used, chip depth changes and the slip plane's magnitude grows in the deformation zone, causing contact between the tool and chip to increase. This makes it necessary to use the cooling/lubrication system in the appropriate conditions for a better cutting operation. As a result, numerous analytical methods have been proposed to calculate the thermal section cutting tool.

The advantages of cutting fluid use can be counted (Debnath et al., 2014; Lawal, 2013):

- Reduce the cutting temperature in the environment and increase the tool life
- Avoid degradation of the surface by preventing the workpiece from sticking to the cutting tool
- Prevents the use of oils from corroding workbench and workpiece
- Keeps the cutting tool working at the same performance

Cutting fluids used in the cooling/lubrication system during machining is still very important due to positive effects such as cooling, lubrication, and cleaning in the machining technique. Due to the toxic effect of cutting fluids used during machining, it is a subject that has occurred in recent years in terms of both environmental factors and worker health. It creates an allergic reaction to skin contact of workers because of the substances contained in cutting fluids and causes infection (Shashidhara and Jayaram, 2010). Petroleum-based cutting fluids containing heterosicism and polyaromatic rings are carcinogenic, according to the International Cancer Research Agency, and exposure may result in professional skin cancer (Abdalla et al., 2007). The use of toxic and biodegradable cutting fluids generates several environmental issues as well as serious health issues among employees (Such as respiratory ailments, lung cancer, and dermatological and genetic diseases) (Cetin et al., 2011). In addition, biotech additives are added to the cutting fluids to prevent bacterial reproduction in cutting fluids. Most biotechnics has also released formaldehyde, which is carcinogenic, so many authorized institutions have banned them from being used. Thus, the disposal of cutting fluids negatively impacts the manufacturer's health and environment, as well as a loss of cost and time (Shokrani et al., 2012). In addition, the use of cutting fluids in machining operations can pose a significant health hazard to machine operators and surrounding personnel. The high pressure and temperature of machining operations can cause cutting fluids to evaporate and atomize, releasing chemical additives such as sulfur, chlorine, phosphorus, hydrocarbons, and biocides into the air. These chemicals can cause respiratory problems and skin reactions, and some of them have been identified as potential carcinogens. Therefore, it is essential to carefully select cutting fluids that have low toxicity and hazard potential, and to implement appropriate safety measures and protective equipment to minimize the risks associated with their use. Additionally, the development of environmentally friendly cutting fluids is a growing area of research and innovation in the field of machining technology (Kalpakjian and Schmid, 2009). As a result, academic research is required to create technologies such as dry, the minimal quantity of lubrication (MQL), the nano-particle additive MQL (NanoMQL), and cryogenic machining.

6.3.1.2.1 Dry Machining

Dry machining does not need the use of cutting fluid. As a result, the heat removed reduces, causing the tool and workpiece to heat up (Weinert et al., 2004). To boost wear resistance, the right tool material selection must be considered while applying dry treatment (El-Bouri, 2018). The tool employed for machining operations necessitates exceptional physical and chemical properties, such as high hardness, resistance to elevated pressures and temperatures, elevated thermal fatigue limit, and excellent chemical stability. Furthermore, the tool must exhibit high toxicity to enable machining operations in various environments (Goindi and Sarkar, 2017).

Dry machining can be effective at lower cutting speeds and when the workpiece does not require precise size and form sensitivity that is influenced by temperature (Sampaio et al., 2018). However, it is important to ensure that hot chips are quickly and effectively removed from the cutting area, and that heat is dissipated from the tool table components (Viswanathan et al., 2018). The key benefit of dry machining is that it does not contaminate the environment or water. Additionally, there is no residue left on the workpiece or chips, which eliminates the need for cleaning and waste liquid disposal, resulting in cost and energy savings.

Although dry machining has its advantages, it also has certain disadvantages such as material sticking to the cutting tool, excessive heat buildup, increased friction between the tool and the workpiece, and inadequate chip removal (Sharma et al., 2016). The dry machining method is used to prevent fire and fire hazards due to the formation of hydrogen caused by the magnesium and water reaction during the treatment of magnesium (Selvam and Sivaram, 2020). Although dry machining has been effective in various machining processes such as magnesium and aluminum, cutting fluids are now more commonly used in high-speed titanium machining (Dargusch et al., 2018; Haddag et al., 2016; Pattnaik et al., 2018; Trujillo Vilches et al., 2017).

6.3.1.2.2 Minimum Quantity Lubrication (MQL)

An innovative, ecologically friendly, and cost-effective approach of minimizing lubrication applications for wet or dry cutting fluid application (Gaurav et al., 2020). This cooling method is also known as dry lubrication or micro-lubrication. Figure 6.6 provides a schematic representation for turning the MQL system (Cagan et al., 2023).

MQL machining is considered a practical way of producing clean, sustainable production (Boswell et al., 2017; Sharma et al., 2015). The key benefits of the MQL are cutting fluid consumption, reducing costs, reducing environmental impact, and improving overall performance in the cutting process and surface quality (Conger et al., 2019). A little quantity of oil/coolant is combined with air to form an aerosol spray into the cutting region with a high-pressure nozzle in the MQL process. This system comprises an atomizer, a cutting fluid carrier, and a discharge nozzle. The atomizer injects the refrigerant with high-pressure air and atomizes it. The resulting atomized cooling is then delivered to the air conditioning plant via a low-pressure distribution system. The cutting fluid is absorbed into the oil reservoir by the venturi effect in the mixing chamber, where it is held at a constant hydraulic load. The

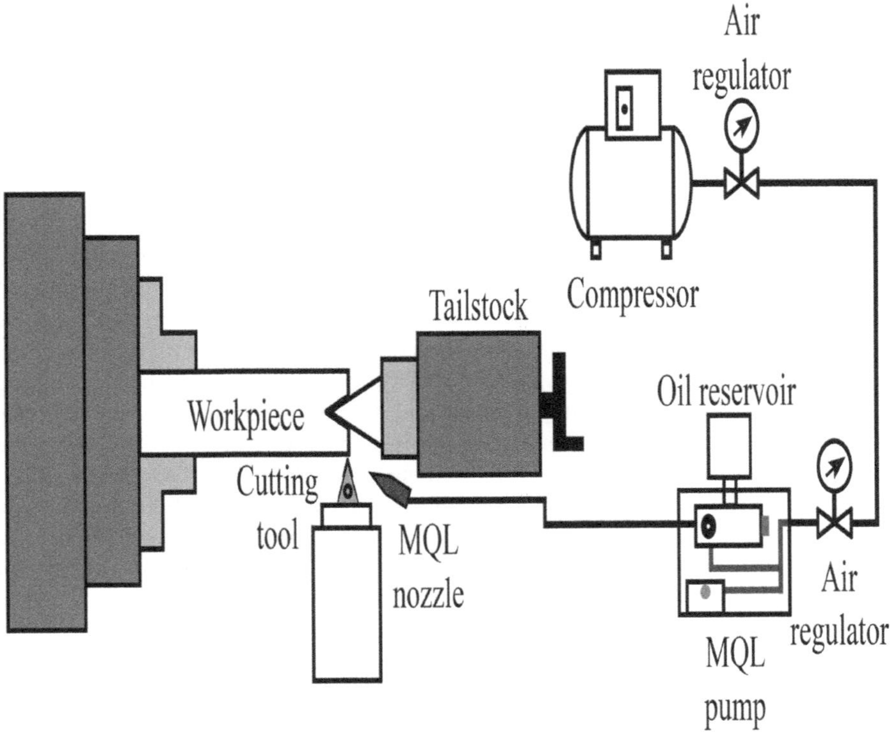

FIGURE 6.6   Schematic representation of the MQL system (Cagan et al., 2023)

coolant flow is then atomized into micron-sized particles in the air flowing through the mixing chamber. When this aerosol cutting tool is sprayed, it acts as a lubricant and cooling medium, penetrating the depths of the cutting tool-work piece interface (Attanasio et al., 2006). The use of cutting fluid in machinability applications varies. Lubrication/cooling systems are used as internal spraying through the cutting tool or spraying directly onto the material cutting tool interface from the outside.

6.3.1.2.3 Minimum Quantity Lubrication with Nano-Particle Additive (Nano MQL)

In recent years, nanofluidics' use has been significantly attracted by its multi-purpose functions in metal machining processes. In particular, it is proven by research that nanofluidics is efficient for their unusual properties in heat transfer and lubrication. The study by Singh and his friends (Singh et al., 2018) indicates that these two characteristics constitute approximately 15%–20% of the total cost of manufacturing a product. However, due to the complex manufacturing phases associated with the synthesis of nano-particles embedded in metal machining fluids, it should be noted that nanofluids have significantly higher production costs than traditional cutting fluids (Das et al., 2021). Nano-particles should be objects with at least one linear size smaller than 100 nm, according to the definition generally accepted in the physics community. As it is known, the production of such materials has revealed some risk management and security control issues arising from the ease of distribution in the environment and living organisms, but with the development of technology, both these environmental and health issues are tried to be avoided, and the cost reduction of the manufacturing industry is seen to increase by day (Kishawy et al., 2019).

The MQL system is a beneficial method for sustainable manufacturing due to its positive contribution to more environmentally friendly human health, although it is superior in the process of metal cutting than traditional cutting fluids (Eltaggaz et al., 2018). However, the MQL system is limited when machining complex processed materials in certain cutting conditions. One of the reasons for this is that the amount of cutting fluid is very little used because it is green. This causes the part not to be sufficiently cooled during the cutting process (Madhukar et al., 2016). It is, therefore, necessary to improve the MQL system's performance, especially when machining certain materials. Recently, in nanometer and micron, metal, metal-oxide, carbide, nitrite, carbon nanotube, and Nano-fluid cutting fluids were obtained by joining the cutting fluids used in the MQL system of many materials such as TiO_2, Al_2O_3, MoS_2 and graphite (Duc et al., 2020; Gupta et al., 2016; Hegab et al., 2020; Singh et al., 2017). Thanks to the nano additives, it is stated that the thermal conductivity and lubrication effect of cutting fluids are increased, thus improving the performance of both environmental and sustainable MQL methods (Kishawy et al., 2019).

6.3.1.3 Environmental Effects of Coolant Functions and Strategies

After reviewing the literature, it was decided that machining dry, MQL, NanoMQL, and cryogenic cooling/lubrication systems will greatly contribute to cleaner and more ecologically responsible manufacturing (Kaynak, 2014; Kaynak et al., 2014). It is also critical to understand which procedure to use for successful machining (i.e., operating information and experienced operators in the workplace). For this reason, machining processes in environments obtained through different cooling/lubrication systems are determined by suitable cutting tool material, workpiece, and cutting conditions. Depending on the cutting conditions, the surface integrity of the part to be obtained is determined. In general, the MQL and NanoMQL do not have any apparent effect on the cutting forces compared to the dry cutting process but significantly improve surface quality. The following statements may help compare dry with MQL and NanoMQL (Kaynak, 2014; Khatri and Jahan, 2018):

- Dry environment occurs when the minimal quantity of lubrication used is close to zero.
- In dry environment, the cutting tool temperature is extremely high, resulting in increased tool wear, lower tool life, and poor machining.
- In dry environment, chips are wrapped around the cutting regions, which harms the treated surface.
- Regardless of how little cutting fluid is used in MQL machining, it pollutes the water, land, and environment very little.
- To reap the greatest benefits from both sustainable approaches, dry and MQL are optimized to eliminate cutting fluid.

One of the main tasks of engineers is to determine the appropriate parameters. The cutting environment differs depending on the material being processed, the surface quality, and the cutting tool tips employed. In this respect, engineers are precious in the combination of theoretical and practical for operators to act with their knowledge and experience.

6.4 CASE STUDIES

6.4.1 Case 1

Khanna et al. (2022a) produced more ecological and energy-efficient industrial techniques and materials in recent years, as citizens' interest about environmental preservation has grown. For this reason, the use of GFRP materials in automotive and aircraft industries, with a lighter, superior specific force, rather than traditional metal alloys, with resistance to corrosion, fatigue, microorganisms, and chemicals, is prominent. An ecologically sustainable cooling system is being investigated as a means of reducing the harmful environmental effects of traditional coolants for machining that are emulsion-based. In this regard, two cooling systems – dry and LCO_2 – were selected, and their performance was evaluated in terms of several various variables, such as life cycle assessments force, tool life, temperature, surface integrity, and power usage. Both of these systems were examined when studying the GFRP milling process. LCO_2 was shown to outperform dry machining, resulting in an 80% decrease in cutting zone temperature, a 5% decrease in cutting force, a 9% decrease in power usage, and a lower level of surface roughness. Nevertheless, compared to dry machining, the compression of CO_2 to liquid form uses more energy and natural resources and has a greater negative impact on the environment.

6.4.2 Case 2

Firmino et al. (2022) determine that companies need to modernize life cycle engineering strategies to gain a competitive advantage with a sustainable production understanding in industry 4.0. As a result, researchers aim to optimize the product lifecycle by integrating machine learning, information management, and other versatile technologies. In this study, equipment usage modes have defined four different machine usage modes according to UPLCI methodology, from starting the production process to switching it off. These are:

- *Initialization:* The machinery sits idle while waiting for the machining cycle to begin.
- *Standby:* The machine has finished a machining cycle and is waiting for finished components to be removed and raw material to be loaded before starting a new cycle.
- *Full load:* Implement the machining procedure. It also moves the tool holder's head, changes tools, and approaches the workpiece after changing tools.
- *Shutdown:* The equipment has completed the daily request and is being turned off.

Six equipment components have been discovered for this metedological using the UPLCI method:

- *Primary system:* Turns on the machine's illumination, hydraulic unit, computers, and machine sensors.
- *Spindle:* The axis of rotation of the machine to which the tool is fastened in order to manufacture.
- *Tool holder:* A holding mechanism created to hold the tools needed for the cycle of the machining
- *Cooling system:* is in charge of delivering coolant (a combination of oil and water) to the tool when it comes into contact with the workpiece.
- *Exhaust system:* This system is in charge of eliminating smoke from the machining process.
- *Chip conveyor:* A continuous system that removes waste chips from the cutting area.

When the findings were examined, these helped to define electricity and fluid usage as the key life cycle hot spots that needed to be reduced. Based on eco-efficiency studies, future production scenarios are advised to lessen lifecycle impacts. The case when each production cycle produces twice as many parts, increases the number of machine components each cycle from three to six, and decreases the raw material volume to be processed by 25% was found to be a "win-win" condition. All of these have been identified as a reduction in lifecycle effects by up to 16.47% and an increase in eco-efficiency by up to 23.55%.

6.4.3 CASE 3

In another study, Khanna et al. (2022b) find that governments have recently increased incentives to improve energy consumption and environmentally friendly machining processes. This study aims to use a more environmentally sustainable system instead of using carbon-based coolants in their machining experiments. Energy consumption of the different cutting fluids and LCA analysis are performed by determining the same machining parameters during the machining of the material. The 15-5 stainless steel material's turning experiments were performed in eight different cutting environments and nine different cutting speeds and feed rates. The different cutting environments are dry, wet, cryogenic spray with LCO_2 gases, cryogenic spray with LN_2 gases, MQL, electrostatic MQL, hybrid (Electrostatic MQL with nano-particles added), and electrostatic lubrication. The dry cutting condition showed the highest energy consumption, while the hybrid condition (Which involves the use of electrostatic MQL with added nanoparticles) showed the lowest energy consumption. Figure 6.7 illustrates the experimental installation of this study. The minimum energy consumption has been achieved in the Hybrid system. Compared to hybrid and other cooling/lubrication systems, energy consumption varies between 4 and 10%. The maximum energy consumption is achieved in a dry environment. Although the hybrid system consumes minimal energy, the nano-particles used in it make it the most environmentally affected system.

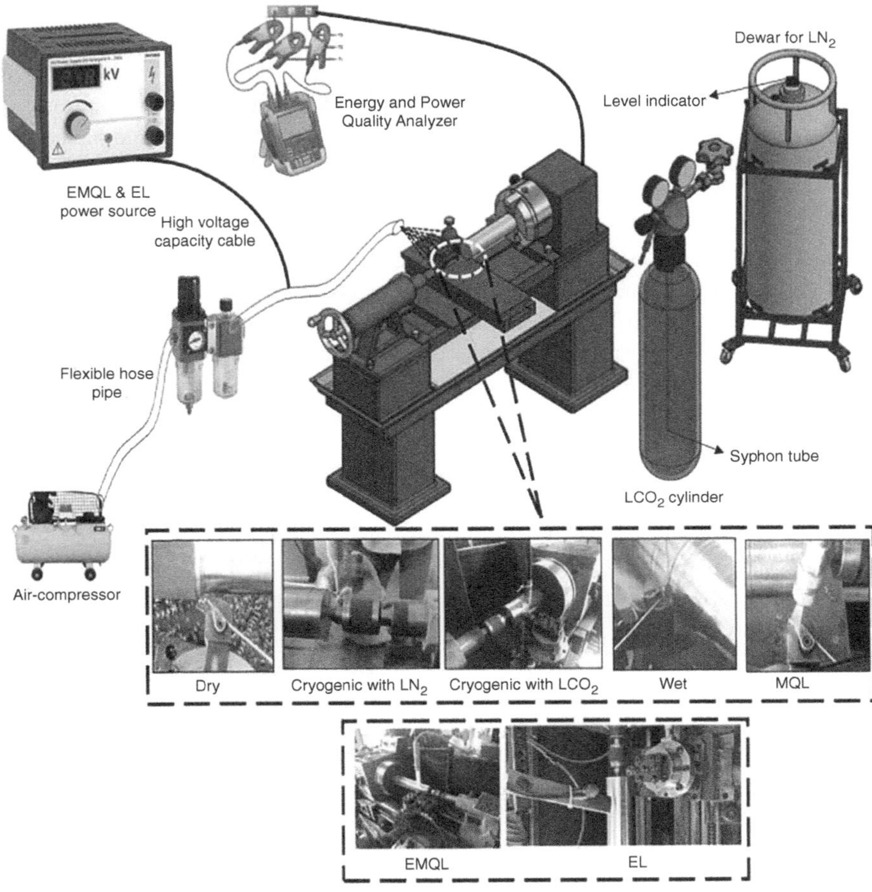

FIGURE 6.7　Experimental installation of the machining process (Khanna et al., 2022b)

6.4.4　CASE 4

The study conducted by Mia et al. (2019) aimed to enhance the machining process of Ti6Al4V alloys in a sustainable and environmentally friendly manner. Due to the high-temperature properties of titanium alloys, achieving sustainable machining targets is challenging. To address this issue, the researchers conducted machining experiments using a cryogenic liquid nitrogen (LN_2) environment. The focus was on the use of a cryogenic cooling system with low carbon emissions to achieve clean machining. The study examined the effects of LN_2 cooling on force, temperature, specific energy, and surface quality, both as a single jet and dual jets. The research was conducted in three parts: first, studying the impact of different variables on output parameters; second, optimizing the parameters for maximum performance; and third, conducting a life cycle analysis of the cryogenic LN_2-assisted machining process. According to the study, the dual-jet LN_2 cooling system was the most efficient in lowering temperature and specific energy use while enhancing surface quality. This is because LN_2 efficiently and rapidly removes heat from key surfaces. Furthermore, the results of the LCA study revealed that environmental factors have a direct influence on the Ti6Al4V alloy's machining performance in various cutting conditions, including those involving resources, energy, human life, and habitat.

6.4.5 CASE 5

Gupta et al. (2020b) have experimented with cutting the Ti grade 2 alloy, which is a difficult-to-process material, using the more advanced version of the MQL system, the Ranque-Hilsch Vortex Tube-assisted MQL cutting fluids. This RHVT-MQL lubrication system has emerged as a sustainable, environmentally viable alternative for the manufacturing industry. MQL systems have indicated that they are preferred because they are more environmentally and economically compared to cutting in dry mediums. The effects of different speeds, feed ratios, and cutting depths on force, power consumption, specific cutting energy, chip types, and their consequences on material removal rate and roughness were investigated in turning operations. Also, an LCA model was created by evaluating sustainability performance using the Simapro 8.3 software. According to the findings, RHVT-MQL conditions performed better than MQL circumstances with regard to machining characteristics under comparable settings.

6.4.6 CASE 6

In this study by Singh and Sharma (2022), Hastelloy C-276 investigated the machining of a superalloy material. This material is nickel-chrome-based and is a difficult-to-cut type compared to other materials. Therefore, they used ultrasonic atomized cutting fluid to improve the machining index, aiming to develop various cutting environments. This results in a suitable cooling/lubrication environment with the aid of tiny atomic droplets. The improved cooling/lubrication environment has improved machining performance and sustainability results compared to other environments. Compared to dry medium, the main force has been 8.8%, the forward force is 11.7% and the radial force is 10.8% reduced. An improvement of approximately 20% in surface roughness and friction coefficient has been observed. Furthermore, the Simapro 8.1 software was used to complete the life cycle analysis (LCA) sustainability assessment.

6.4.7 CASE 7

In the research conducted by Pereira et al. (2016), the study aimed to investigate the performance of AISI 304 SS under varied turning settings and cutting environments in terms of productiveness, machining efficiency, and environmental friendliness. The experiment examined the impact of different cutting environments such as dry, MQL, LN_2, CO_2, CryoMQL-LN_2, and CryoMQL-CO_2 on the cutting force, dimensional accuracy, and the tool geometry. When the data were examined, the CryoMQL-CO_2 environment improved tool life by more than 50% compared to dry machining, and the cutting speed increased by more than 30%. Furthermore, the condition changed when force and roughness were studied. A lifespan assessment has been performed from an ecological standpoint, allowing comparison of the several proposed solutions. They used six different assumptions in this LCA. These are the interpretations of cutting tool, cutting fluid leaks, cutting fluid operations, CO_2 and cutting fluids, power consumption, and data. Removing this heat from the center due to the heat released during stainless steel machining will increase the efficiency of the machining. This is also possible with an advanced cooling/lubrication system. Dry treatment is not economically and sustainably applicable, as it causes tool wear. The LCA analysis concluded that cryogenic machining and MQL processes are more environmentally friendly than traditional wet machining. Combining cryogenic and MQL techniques is key to balancing technical and environmental issues.

6.4.8 CASE 8

Karim et al. (2021) aim to develop a sustainable machining method that is green, economical, and technically applicable, unlike traditional machinability. Al-Mg-Zr exposed an alloy material to machining in different cutting environments (Dry and MQL) using a diamond polycrystal cutting tool and examined its effects on the roughness and temperature. In addition, these results were evaluated with analysis methods such as ANOVA, ANN and LCA. It was concluded that MQL-assisted machining is a sustainable process compared to dry machining, with very little use of the cutting fluid and a smaller carbon footprint. By employing MQL system in machining operations, the temperature at the tool interface is decreased, which leads to decreased energy consumption and lower tool wear while removing material. This saves time, energy, and cost and provides green and sustainable machining. However, the use of this method in industrial applications provides 11 data to increase sustainable manufacturing.

6.5 SUMMARY

In this book chapter, studies have been examined on the use of the LCE approach in machining, sustainability, LCE approach, and information about the machining process. In today's era, the manufacturing industry is facing increasing pressure from various regulations and policies implemented by developed and developing countries to mitigate the adverse effects of industrial activities on the environment and global climate. The implementation of these measures is crucial to ensure sustainable development and preserve the planet's ecological balance for present and future generations. Companies that adopt the principle of cradle-to-grave as the main objective of the LCE are sustainable and environmentally responsible, so they are moving forward in a competitive market as a reason for choice by customers. Therefore, developing the LCE approach in every field has become a must for our world. In this section, the use of the LCE approach for environmental and sustainable manufacturing in machining applications is increasing in light of the data obtained from the study. In the machining of challenging materials, conventional practices usually involve the application of cutting fluids, utilization of cutting machinery, and consumption of energy. The main effort of engineers is to minimize the economic, occupational and occupational health, and environmental impact in the first place. The primary goal for improving the use of these materials in the industry has been identified as identifying the appropriate cutting parameters and improving more environmentally friendly cutting environments. Researchers aim to achieve sustainable environmental manufacturing by performing the appropriate cutting parameters in the appropriate cutting environment.

REFERENCES

Abdalla, H. S., Baines, W., & Slade, C. (2007). Development of novel sustainable neat-oil metal working fluids for stainless steel and titanium alloy machining. Part 1. Formulation development. *The International Journal of Advanced Manufacturing Technology*, 34(1–2), 21–33.

Alting, L., Hauschild, M., & Wenzel, H. (2007). *Life Cycle Engineering and Management: Status and Perspectives Sustainability in Manufacturing: Recovery of Resources in Product and Material Cycles* (pp. 31–67): Springer Science+ Business Media.

Arrazola, P., Özel, T., Umbrello, D., Davies, M., & Jawahir, I. (2013). Recent advances in modelling of metal machining processes. *CIRP Annals*, 62(2), 695–718.

Arvanitoyannis, I. S. (2008). ISO 14040: Life cycle assessment (LCA)–principles and guidelines. *Waste Management for the Food Industries* (pp. 97–132). Elsevier Academic Press.

Attanasio, A., Gelfi, M., Giardini, C., & Remino, C. (2006). Minimal quantity lubrication in turning: Effect on tool wear. *Wear*, 260(3), 333–338.

Benedicto, E., Carou, D., & Rubio, E. (2017). Technical, economic and environmental review of the lubrication/cooling systems used in machining processes. *Procedia Engineering*, 184, 99–116.

Bojarska, J., Złoty, P., & Wolf, W. M. (2021). Life cycle assessment as tool for realization of sustainable development goals-towards sustainable future of the world: Mini review. *Acta Innovations*, 38, 49–61.

Borowy, I. (2013). *Defining Sustainable Development for Our Common Future: A History of the World Commission on Environment and Development (Brundtland Commission)*: Routledge.

Boswell, B., Islam, M. N., Davies, I. J., Ginting, Y., & Ong, A. K. (2017). A review identifying the effectiveness of minimum quantity lubrication (MQL) during conventional machining. *The International Journal of Advanced Manufacturing Technology*, 92(1), 321–340.

Cagan, S. C., Tasci, U., Pruncu, C. I., & Bostan, B. (2023). Investigation of the effects of eco-friendly MQL system to improve the mechanical performance of WE43 magnesium alloys by the burnishing process. *Journal of the Brazilian Society of Mechanical Sciences and Engineering*, 45(1), 22.

Cetin, M. H., Ozcelik, B., Kuram, E., & Demirbas, E. (2011). Evaluation of vegetable based cutting fluids with extreme pressure and cutting parameters in turning of AISI 304L by Taguchi method. *Journal of Cleaner Production*, 19(17–18), 2049–2056.

Conger, D., Emiroglu, U., & Altan, E. (2019). An experimental study on cutting forces and surface roughness in MQL milling of aluminum 6061. *Machines Technologies Materials*, 13(2), 86–89.

Curran, M. A. (2012). *Life Cycle Assessment Handbook: A Guide for Environmentally Sustainable Products*: John Wiley & Sons.

Dargusch, M. S., Sun, S., Kim, J. W., Li, T., Trimby, P., & Cairney, J. (2018). Effect of tool wear evolution on chip formation during dry machining of Ti-6Al-4V alloy. *International Journal of Machine Tools and Manufacture*, 126, 13–17.

Das, A., Patel, S. K., Arakha, M., Dey, A., & Biswal, B. B. (2021). Processing of hardened steel by MQL technique using nano cutting fluids. *Materials and Manufacturing Processes*, 36(3), 316–328.

Davim, J. P. (2008). *Machining: Fundamentals and Recent Advances:* Springer.

Debnath, S., Reddy, M. M., & Yi, Q. S. (2014). Environmental friendly cutting fluids and cooling techniques in machining: A review. *Journal of Cleaner Production*, 83, 33–47.

Denis, C. (2005). *McGraw-Hill Machining and Metalworking Handbook (McGraw-Hill Handbooks)*: McGraw-Hill Professional.

Duc, T. M., Long, T. T., & Van Thanh, D. (2020). Evaluation of minimum quantity lubrication and minimum quantity cooling lubrication performance in hard drilling of Hardox 500 steel using Al_2O_3 nanofluid. *Advances in Mechanical Engineering*, 12(2), 1687814019888404.

El-Bouri, W. (2018). *Investigation of Minimum Quantity Lubrication Coolant Strategy for the Machining of Austempered Ductile Iron (ADI)*: University of Guelph.

El-Hofy, H. (2018). *Fundamentals of Machining Processes: Conventional and Nonconventional Processes*: CRC Press.

Eltaggaz, A., Hegab, H., Deiab, I., & Kishawy, H. (2018). Hybrid nano-fluid-minimum quantity lubrication strategy for machining austempered ductile iron (ADI). *International Journal on Interactive Design and Manufacturing (IJIDeM)*, 12(4), 1273–1281.

Faruk, O., Tjong, J., & Sain, M. (2017). *Lightweight and Sustainable Materials for Automotive Applications*: CRC Press, Boca Raton.

Finkbeiner, M., Inaba, A., Tan, R., Christiansen, K., & Klüppel, H.-J. (2006). The new international standards for life cycle assessment: ISO 14040 and ISO 14044. *The International Journal of Life Cycle Assessment*, 11(2), 80–85.

Firmino, A. S., Akim, E. K., de Oliveira, J. A., & Silva, D. A. L. (2022). Green machining of aluminum pipes: An integrated approach for eco-efficiency and life cycle assessment in manufacturing systems. *The International Journal of Advanced Manufacturing Technology*, 121(9), 6225–6241.

Fleischer, J., Pabst, R., & Kelemen, S. (2007). Heat flow simulation for dry machining of power train castings. *CIRP Annals*, 56(1), 117–122.

Garetti, M., & Taisch, M. (2012). Sustainable manufacturing: Trends and research challenges. *Production Planning & Control*, 23(2–3), 83–104.

Gaurav, G., Sharma, A., Dangayach, G., & Meena, M. (2020). Assessment of jojoba as a pure and nano-fluid base oil in minimum quantity lubrication (MQL) hard-turning of Ti–6Al–4V: A step towards sustainable machining. *Journal of Cleaner Production*, 272, 122553.

Goindi, G. S., & Sarkar, P. (2017). Dry machining: A step towards sustainable machining–Challenges and future directions. *Journal of Cleaner Production*, 165, 1557–1571.

Groover, M. P. (2020). *Fundamentals of Modern Manufacturing: Materials, Processes, and Systems*: John Wiley & Sons.

Grote, K.-H., & Antonsson, E. K. (2009). *Springer Handbook of Mechanical Engineering* (Vol. 10): Springer.

Guinee, J. B., Heijungs, R., Huppes, G., Zamagni, A., Masoni, P., Buonamici, R., . . . Rydberg, T. (2011). *Life Cycle Assessment: Past, Present, and Future*: ACS Publications.

Gupta, M. K., Mia, M., Jamil, M., Singh, R., Singla, A. K., Song, Q., . . . Sarikaya, M. (2020a). Machinability investigations of hardened steel with biodegradable oil-based MQL spray system. *The International Journal of Advanced Manufacturing Technology*, 108(3), 735–748.

Gupta, M. K., Song, Q., Liu, Z., Pruncu, C. I., Mia, M., Singh, G., . . . Jamil, M. (2020b). Machining characteristics based life cycle assessment in eco-benign turning of pure titanium alloy. *Journal of Cleaner Production*, 251, 119598.

Gupta, M. K., Sood, P., & Sharma, V. S. (2016). Optimization of machining parameters and cutting fluids during nano-fluid based minimum quantity lubrication turning of titanium alloy by using evolutionary techniques. *Journal of Cleaner Production*, 135, 1276–1288.

Haddag, B., Atlati, S., Nouari, M., & Moufki, A. (2016). Dry machining aeronautical aluminum alloy AA2024-T351: Analysis of cutting forces, chip segmentation and built-up edge formation. *Metals*, 6(9), 197.

Hao, G., & Liu, Z. (2020). The heat partition into cutting tool at tool-chip contact interface during cutting process: A review. *The International Journal of Advanced Manufacturing Technology*, 108(1), 393–411.

Hauschild, M. Z., Kara, S., & Røpke, I. (2020). Absolute sustainability: Challenges to life cycle engineering. *CIRP Annals*, 69(2), 533–553.

Hauschild, M. Z., Rosenbaum, R. K., & Olsen, S. I. (2018). *Life Cycle Assessment*: Springer.

Hegab, H., Umer, U., Esawi, A., & Kishawy, H. (2020). Tribological mechanisms of nano-cutting fluid minimum quantity lubrication: A comparative performance analysis model. *The International Journal of Advanced Manufacturing Technology*, 108(9), 3133–3139.

Herrmann, C., Hauschild, M., Gutowski, T., & Lifset, R. (2014). *Life Cycle Engineering and Sustainable Manufacturing* (Vol. 18, pp. 471–477): Wiley.

Jain, V. K. (2009). *Advanced Machining Processes*: Allied Publishers.

Jamwal, A., Agrawal, R., & Sharma, M. (2021). *Life Cycle Engineering: Past, Present, and Future Sustainable Manufacturing* (pp. 313–338): Elsevier.

Kalpakjian, S., & Schmid, S. R. (2009). *Manufacturing Engineering. Technology* (pp. 661–663): Addison-Wesley, Reading, MA (1989).

Karim, M. R., Tariq, J. B., Morshed, S. M., Shawon, S. H., Hasan, A., Prakash, C., Pruncu, C. I. (2021). Environmental, economical and technological analysis of MQL-assisted machining of Al-Mg-Zr alloy using PCD tool. *Sustainability*, 13(13), 7321.

Kaswan, M. S., Rathi, R., Khanduja, D., & Singh, M. (2020). *Life Cycle Assessment Framework for Sustainable Development in Manufacturing Environment Advances in Intelligent Manufacturing* (pp. 103–113): Springer.

Kaynak, Y. (2014). Evaluation of machining performance in cryogenic machining of Inconel 718 and comparison with dry and MQL machining. *The International Journal of Advanced Manufacturing Technology*, 72(5), 919–933.

Kaynak, Y., Lu, T., & Jawahir, I. (2014). Cryogenic machining-induced surface integrity: A review and comparison with dry, MQL, and flood-cooled machining. *Machining Science and Technology*, 18(2), 149–198.

Khanna, N., Shah, P., Sarikaya, M., & Pusavec, F. (2022a). Energy consumption and ecological analysis of sustainable and conventional cutting fluid strategies in machining 15–5 PHSS. *Sustainable Materials and Technologies*, e00416.

Khanna, N., Rodríguez, A., Shah, P., Pereira, O., Rubio-Mateos, A., de Lacalle, L. N. L., & Ostra, T. (2022b). Comparison of dry and liquid carbon dioxide cutting conditions based on machining performance and life cycle assessment for end milling GFRP. *The International Journal of Advanced Manufacturing Technology*, 122(2), 821–833.

Khatri, A., & Jahan, M. P. (2018). Investigating tool wear mechanisms in machining of Ti-6Al-4V in flood coolant, dry and MQL conditions. *Procedia Manufacturing*, 26, 434–445.

Khoshnevis Yazdi, S., Shakouri, B., Salehi, H., & Fashandi, A. (2017). Sustainable development and ecological economics. *Energy Sources, Part B: Economics, Planning, and Policy*, 12(8), 740–748.

Kishawy, H. A., Hegab, H., Deiab, I., & Eltaggaz, A. (2019). Sustainability assessment during machining Ti-6Al-4V with Nano-additives-based minimum quantity lubrication. *Journal of Manufacturing and Materials Processing*, 3(3), 61.

Klöpffer, W. (2012). The critical review of life cycle assessment studies according to ISO 14040 and 14044. *The International Journal of Life Cycle Assessment*, 17(9), 1087–1093.

Kravchenko, M., Pigosso, D. C., & McAloone, T. C. (2019). Towards the ex-ante sustainability screening of circular economy initiatives in manufacturing companies: Consolidation of leading sustainability-related performance indicators. *Journal of Cleaner Production*, 241, 118318.

Kuhlman, T., & Farrington, J. (2010). What is sustainability? *Sustainability*, 2(11), 3436–3448.

Lawal, S. A. (2013). A Review of Application of Vegetable Oil-Based Cutting Fluids in Machining Non-Ferrous Metals. *Indian Journal of Science and Technology*, 6(1), 3951–3956.

Machado, C. G., et al. (2020). Sustainable manufacturing in Industry 4.0: An emerging research agenda. *International Journal of Production Research*, 58(5), 1462–1484.

Madanchi, N. (2021). *Model Based Approach for Energy and Resource Efficient Machining Systems*: Springer.

Madhukar, S., et al. (2016). A critical review on minimum quantity lubrication (MQL) coolant system for machining operations. *International Journal of Current Engineering and Technology*, 6(5), 1745–1751.

Malek, J., & Desai, T. N. (2020). A systematic literature review to map literature focus of sustainable manufacturing. *Journal of Cleaner Production*, 256, 120345.

Mia, M., Gupta, M. K., Lozano, J. A., Carou, D., Pimenov, D. Y., Królczyk, G., . . . Dhar, N. R. (2019). Multi-objective optimization and life cycle assessment of eco-friendly cryogenic N2 assisted turning of Ti-6Al-4V. *Journal of Cleaner Production*, 210, 121–133.

Mills, B. (2012). *Machinability of Engineering Materials*: Springer Science & Business Media.

Murray, G. (1997). *Handbook of Materials Selection for Engineering Applications*: CRC Press.

Nee, A. Y. C. (2014). *Handbook of Manufacturing Engineering and Technology*: Springer Publishing Company, Incorporated.

Pacheco Rocha, N., Dias, A., Santinha, G., Rodrigues, M., Rodrigues, C., Queirós, A., . . . Pavão, J. (2022). Systematic literature review of context-awareness applications supported by smart cities' infrastructures. *SN Applied Sciences*, 4(4), 1–19.

Panda, A., Das, S. R., & Dhupal, D. (2018). Experimental investigation, modelling and optimization in hard turning of high strength low alloy steel (AISI 4340). *Matériaux & Techniques*, 106(4), 404.

Pattnaik, S. K., Bhoi, N. K., Padhi, S., & Sarangi, S. K. (2018). Dry machining of aluminum for proper selection of cutting tool: Tool performance and tool wear. *The International Journal of Advanced Manufacturing Technology*, 98(1), 55–65.

Penciuc, D., Le Duigou, J., Daaboul, J., Vallet, F., & Eynard, B. (2016). Product life cycle management approach for integration of engineering design and life cycle engineering. *AI EDAM*, 30(4), 379–389.

PennState. (2020). *Technologies for Sustainability Systems*: Pennstate University.

Pereira, O., et al. (2016). Cryogenic and minimum quantity lubrication for an eco-efficiency turning of AISI 304. *Journal of Cleaner Production*, 139, 440–449.

Pimenov, D. Y., Mia, M., Gupta, M. K., Machado, Á. R., Pintaude, G., Unune, D. R., . . . Wojciechowski, S. (2022). Resource saving by optimization and machining environments for sustainable manufacturing: A review and future prospects. *Renewable and Sustainable Energy Reviews*, 166, 112660.

Purvis, B., Mao, Y., & Robinson, D. (2019). Three pillars of sustainability: In search of conceptual origins. *Sustainability Science*, 14(3), 681–695.

Pusavec, F., Krajnik, P., & Kopac, J. (2010). Transitioning to sustainable production–Part I: Application on machining technologies. *Journal of Cleaner Production*, 18(2), 174–184.

Rebitzer, G., Ekvall, T., Frischknecht, R., Hunkeler, D., Norris, G., Rydberg, T., . . . Pennington, D. W. (2004). Life cycle assessment: Part 1: Framework, goal and scope definition, inventory analysis, and applications. *Environment International*, 30(5), 701–720.

Redclift, M. (2005). Sustainable development (1987–2005): An oxymoron comes of age. *Sustainable Development*, 13(4), 212–227.

Sampaio, M. A., Machado, Á. R., Laurindo, C. A. H., Torres, R. D., & Amorim, F. L. (2018). Influence of minimum quantity of lubrication (MQL) when turning hardened SAE 1045 steel: A comparison with dry machining. *The International Journal of Advanced Manufacturing Technology*, 98(1), 959–968.

Sangwan, K. S., & Mittal, V. K. (2015). A bibliometric analysis of green manufacturing and similar frameworks. *Management of Environmental Quality: An International Journal*, 26(4), 566–587.

Santos, J., Gouveia, R. M., & Silva, F. (2017). Designing a new sustainable approach to the change for lightweight materials in structural components used in truck industry. *Journal of Cleaner Production*, 164, 115–123.

Selvam, M. D., & Sivaram, N. (2020). A comparative study on the surface finish achieved during turning operation of AISI 4340 steel in flooded, near-dry and dry conditions. *Australian Journal of Mechanical Engineering*, 18(3), 457–466.

Shaked, S., Crettaz, P., Saade-Sbeih, M., Jolliet, O., & Jolliet, A. (2015). *Environmental Life Cycle Assessment*: CRC Press.

Sharma, A. K., Tiwari, A. K., & Dixit, A. R. (2016). Effects of Minimum Quantity Lubrication (MQL) in machining processes using conventional and nanofluid based cutting fluids: A comprehensive review. *Journal of Cleaner Production*, 127, 1–18.

Sharma, V. S., Singh, G., & Sørby, K. (2015). A review on minimum quantity lubrication for machining processes. *Materials and Manufacturing Processes*, 30(8), 935–953.

Shashidhara, Y., & Jayaram, S. (2010). Vegetable oils as a potential cutting fluid—an evolution. *Tribology International*, 43(5–6), 1073–1081.

Shokrani, A., & Biermann, D. (2020). *Advanced Manufacturing and Machining Processes* (Vol. 4, p. 102): MDPI.

Shokrani, A., Dhokia, V., & Newman, S. T. (2012). Environmentally conscious machining of difficult-to-machine materials with regard to cutting fluids. *International Journal of Machine Tools and Manufacture*, 57, 83–101.

Silva, D. A., Filleti, R. A., Christoforo, A. L., Silva, E. J., & Ometto, A. R. (2015). Application of life cycle assessment (LCA) and design of experiments (DOE) to the monitoring and control of a grinding process. *Procedia CIRP*, 29, 508–513.

Singh, R., & Sharma, V. (2022). Experimental investigations into sustainable machining of Hastelloy C-276 under different lubricating strategies. *Journal of Manufacturing Processes*, 75, 138–153.

Singh, R. K., Sharma, A. K., Dixit, A. R., Tiwari, A. K., Pramanik, A., & Mandal, A. (2017). Performance evaluation of alumina-graphene hybrid nano-cutting fluid in hard turning. *Journal of Cleaner Production*, 162, 830–845.

Singh, T., Dureja, J. S., Dogra, M., & Bhatti, M. S. (2018). Environment friendly machining of Inconel 625 under nano-fluid minimum quantity lubrication (NMQL). *International Journal of Precision Engineering and Manufacturing*, 19(11), 1689–1697.

Smith, G. T. (2008). *Cutting Tool Technology: Industrial Handbook*: Springer Science & Business Media.

Sonnemann, G., & Margni, M. (2015). *Life Cycle Management*: Springer Nature.

Taylor, F. W. (1906). *On the Art of Cutting Metals* (Vol. 23): American Society of Mechanical Engineers.

Trujillo Vilches, F. J., et al. (2017). Analysis of the chip geometry in dry machining of aeronautical aluminum alloys. *Applied Sciences*, 7(2), 132.

Viswanathan, R., Ramesh, S., & Subburam, V. (2018). Measurement and optimization of performance characteristics in turning of Mg alloy under dry and MQL conditions. *Measurement*, 120, 107–113.

Weinert, K., Inasaki, I., Sutherland, J., & Wakabayashi, T. (2004). Dry machining and minimum quantity lubrication. *CIRP Annals*, 53(2), 511–537.

Youssef, H., & El-Hofy, H. (2020). *Machining Technology and Operations: 2-Volume Set*: CRC Press.

Youssef, H., & El-Hofy, H. (2022). *Machining Technology and Operations: 2-Volume Set*: CRC Press.

7 Sustainable Machining of Difficult-to-Machine Materials

Emine Şirin

7.1 INTRODUCTION

Today, high-productivity, high-quality, low-cost, and eco-friendly machining are essential starting points in the manufacturing industry. For this purpose, studies are carried out on new production techniques that improve machining performance and product quality. Especially the cooling systems used in the manufacturing industry must be efficient, economical, and suitable for employee and environmental health (Umbrello et al., 2012). Ecological concerns and employee health are critical in the Green Machining culture. Various cooling methods are used to increase the tool life used in metal cutting operations and obtain the desired surface integrity. It is preferred that these cooling methods be suitable for employee and environmental health and economic health. Cooling methods used in the sustainable machining culture aim to minimize or reduce the adverse effects of conventional fluids regarding employee health and environmental concerns. Dry, minimal quantity lubrication (MQL), nanoparticle additions (NanoMQL), and cryogenic cooling using cryogen gases are ecological cooling/lubrication technique samples.

Dry cutting is an efficient method in metalworking because of the total costs and ecological concerns. Nevertheless, dry machining falls short regarding machine efficiency, particularly under heavy machining conditions. Dry machining is, therefore, not highly preferred in terms of machining performance. The MQL method, often called semi-dry machining, reduces or eliminates the adverse effects of dry cutting. Since minimal cutting oils were used in the MQL method, it is preferred for reducing human health and ecological concerns. On the other hand, to enhance the cutting oil's success and tribological effects in heavy machining conditions, nano-sized (typically 1–100 nm) solid particles are doped into the oil (Ghadimi et al., 2011). By adding nanoparticles into the cutting zone under high pressure and at different mixing ratios, the oil mixture's lubricating properties are enhanced, its thermal conductivity is raised, and a thin oil film layer is created between the cutting tool and the material interface (Sharma et al., 2015a; Xie et al., 2016). Nanofluids can produce the required results by preventing the high temperature produced during the cutting process from penetrating the workpiece, lowering the cutting force and tool wear, improving the material surface quality, and extending the tool life (Şirin and Kıvak, 2021; Yıldırım et al., 2019).

DOI: 10.1201/9781003352402-7

Another preferred cooling method in sustainable machining operations is cryogenic machining. Cryogens sent to the cutting area rapidly cool the cutting area, providing an efficient machining process. Nitrogen (N), carbon dioxide (CO_2), helium (He), etc., cryogens are preferred as coolants in cryogenic processing. The gases used as refrigerants evaporate harmlessly in the atmosphere.

Recent years have seen the popularity of hybrid cooling/lubrication technologies (Cryo + MQL, Cryo + NanoMQL), which offer both lubrication and cooling simultaneously. The researchers continue to study the environmental hybrid cooling/lubrication method, which has recently been used in the literature.

The environmentally friendly cooling lubrication techniques for sustainable machining are focused. For this purpose, dry Machining, MQL machining, nanofluid Machining, cryogenic Machining, and hybrid cooling/lubrication machining methods are investigated. The readers were provided knowledge regarding sustainable or Green Machining cooling/lubrication methods in this chapter, and examples of hard-to-cut materials were shown using these cooling/lubrication methods.

7.2 SUSTAINABLE OR GREEN MACHINING TECHNIQUES

At the end of the 21st century, the positive effects of technological developments such as quality, improved efficiency, time-saving, and high productivity have significantly affected human life. Even with these positive results, adverse environmental and human health impacts have worsened. The increase in adverse effects also raised some concerns, especially in the manufacturing industry. The role of the manufacturing industry in economic growth is excellent. Due to rapid production techniques, the factors causing environmental pollution have increased, and new needs have arisen (Gupta, 2020). Companies in the manufacturing industry and ecological

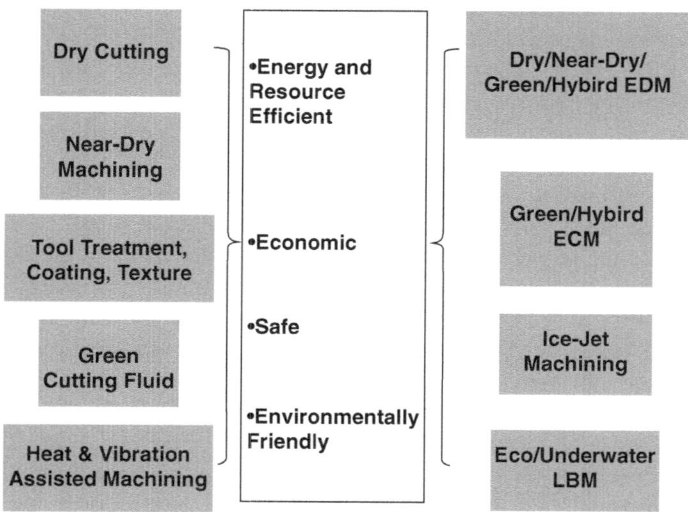

FIGURE 7.1   Green techniques in conventional and advanced machining (Gupta, 2020)

organizations began to use the term "sustainable machining" and "green machining" to meet new expectations. Green machining is advantageous for the ecology and will be seen as essential by manufacturing organizations shortly (Jain and Kansal, 2017) (Figure 7.1).

Metal-cutting fluids increase processing efficiency in metalworking processes (Haider and Hashmi, 2014). In manufacturing processes, i.e., turning, milling, drilling, and grinding, cutting tools account for approximately 4% of production costs, and metal cutting fluids for about 17%. Yet, coolant expenses might reach as high as 20%–30% of production costs when functioning under demanding machining situations and on hard-cutting materials (Pusavec et al., 2010). Traditional fluid provides three essential functions: lubricating, cooling, and removing chips from the cutting area. Indeed, cutting fluids greatly lower the temperature of the material and the tool, improving the surface quality and prolonging the tool's life (Şirin and Kıvak, 2019). However, cutting fluids negatively impact the environment, the employees' health, and the production cost (Anandan et al., 2021). Researchers and producers have concentrated recently on finding ways to utilize less cutting fluid without affecting production efficiency. These techniques are environmentally friendly and extremely important for sustainable manufacturing. Green Machining topics include dry machining, MQL machining, nanoMQL machining, cryogenic cooling machining, hybrid cooling machining, and high-pressure cooling machining.

7.2.1 DRY MACHINING

Unlubricated or uncooled machining is known as dry machining. Due to the protection of environmental and health standards, industries choose dry processing. The benefits of dry machining are: it does not harm the atmosphere (or water), it does not require disposal costs, there is no residue on the chips, and dry machining is not threatening to the environment and worker health and is non-allergenic. During dry machining, high heat, and friction are produced at the tool-workpiece interfaces. However, dry machining is not preferred, especially under heavy machining conditions, since its processing efficiency is very low (Haider and Hashmi, 2014; Sreejith and Ngoi, 2000) (Figure 7.2).

7.2.1.1 Dry Machining Case Study 1

Devillez et al. turned Inconel 718 superalloy material in dry and fluid conventional cooling conditions. At the study's conclusion, the authors tried to compare the surface integrity values of the cost-effective and environmentally friendly dry machining condition with wet cutting. The experimental investigation resulted in the authors concluding that in terms of machining performance and surface integrity, dry cutting without coolants or lubricants was comparable to wet cutting (Devillez et al., 2011).

7.2.2 MQL MACHINING

In industrial settings, the MQL technique, also known as nearly entirely machining, is often used (Sartori et al., 2018). MQL is performed by spritzing the cutting

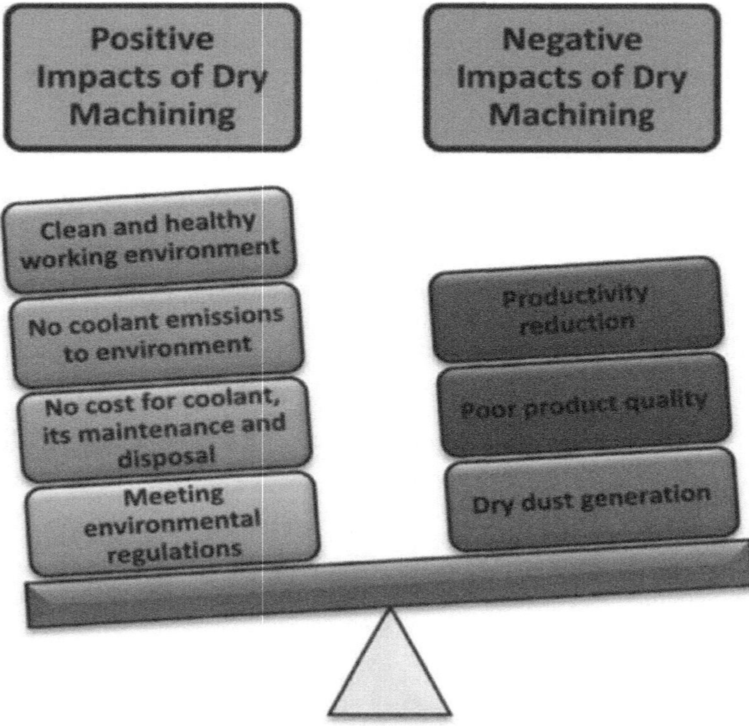

FIGURE 7.2 A balance between the advantages and disadvantages of dry machining (Haider and Hashmi, 2014)

area with a solution of air and a small number of biodegradable fluid particles. The MQL is a method in which a minimal cutting fluid (20 mL/h) combines with pressurized air to form an aerosol mist and is sent directly to the cutting area via a nozzle. With the MQL technique, the cutting fluid passes the cutting region directly and creates a beneficial oil film layer. This oil film layer effectively lowers the cutting temperature and reduces tool wear. Due to its biodegradability and lack of toxic components, the oil used in the MQL process faces a slight risk to the environment or the workers' health. Additionally, MQL is a substitute for dry machining that lowers cutting forces to prolong the tool's life and improve the workpiece's surface quality (Mia et al., 2017). MQL can be used efficiently in machining processes such as milling, turning, drilling, and grinding (Fratila and Production, 2011).

In the MQL method, the compressed oil mixture is applied to the cutting region form aerosol mist in two ways: inside or outside the cutting tool.

In the external MQL method, lubrication is provided in two ways, single and dual channels. With the single-channel MQL approach used external to the cutting tool, compressed air and oil are premixed and transferred to the cutting region via one or

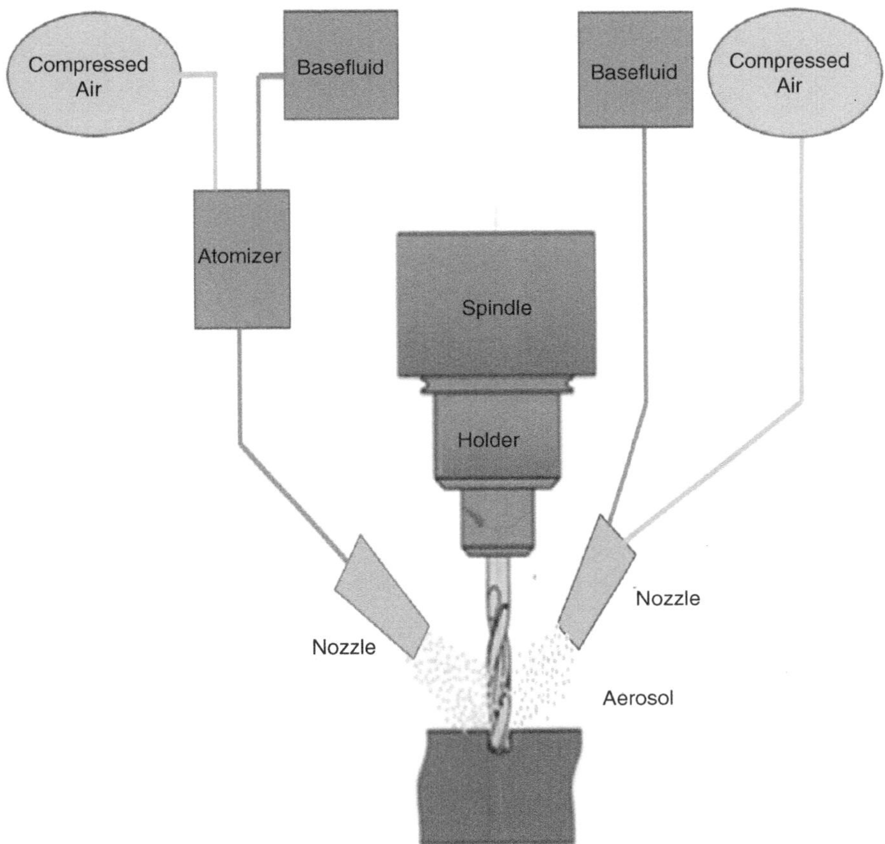

FIGURE 7.3 External MQL methods

more nozzles. In double-channel MQL applications, oil and air come into the nozzle through different channels, are mixed in the nozzle, and sent to the cutting area (Figure 7.3).

The oil-air combination gets to the cutting region via the proper channels that flow through the cutting tool. Compressed air provides cooling, and oil particles offer lubrication in the internal MQL method applied through the cutting tool. There are two methods for applying the internal MQL using the cutting tool: single and double channels. Oil and air are premixed and sent to the cutting area in single-channel applications. Oil and air are transported through separate channels in double-channel applications, and after being mixed in the cutting tool, the mixture is transferred to the cutting region (Figure 7.4).

The nozzle's angle and position are essential in machining efficiency. The experimental set of the Nano MQL application is presented in Figure 7.5.

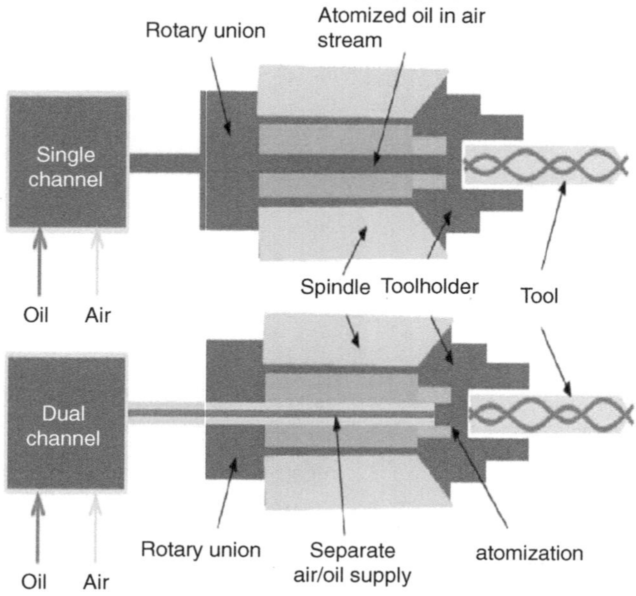

FIGURE 7.4   Internal MQL methods (Tai et al., 2014)

7.2.2.1   MQL Machining Case Study 1

Priarone et al. milled and turned titanium aluminides alloy using dry, wet, and MQL cooling/lubrication procedures. The authors stated that the MQL approach demonstrated good improvements in surface roughness and tool wear based on the findings of the experimental investigation (Figures 7.6 and 7.7) (Priarone et al., 2014).

7.2.3   NanoMQL Machining

The MQL technique is insufficient in terms of machining efficiency due to the advancement of material technologies, the integration of innovative materials, the development of processing, and extreme heat generation. In this case, the need to develop of MQL method has emerged. Hence, the utilization of nanofluids has started, which are created by combining solid nanoparticles into cutting fluids.

Nanofluids are obtained by mixing solid particles in nanometer (nm) size (usually <100) with water, ethylene glycol (EG), synthetic oil, vegetable oil, etc., and base liquid. The nanofluids mixtures were prepared with nanoparticles such as metallic nanoparticles: copper (Cu), nickel (Ni), alumina (Al), silver (Ag), titanium metal oxide: titanium dioxide (TiO_2), zirconium dioxide (ZrO_2), barium titanate ($BaTiO_3$), aluminum oxide (Al_2O_3), copper oxide (CuO), iron oxide (Fe_3O_4), silicon dioxide (SiO_2) and zinc oxide (ZnO), and as a compound: silicon carbide (SiC), carbon nanotube (CNT), aluminum nitride (AlN), graphene, and calcium carbonate ($CaCO_3$) (Das, 2017).

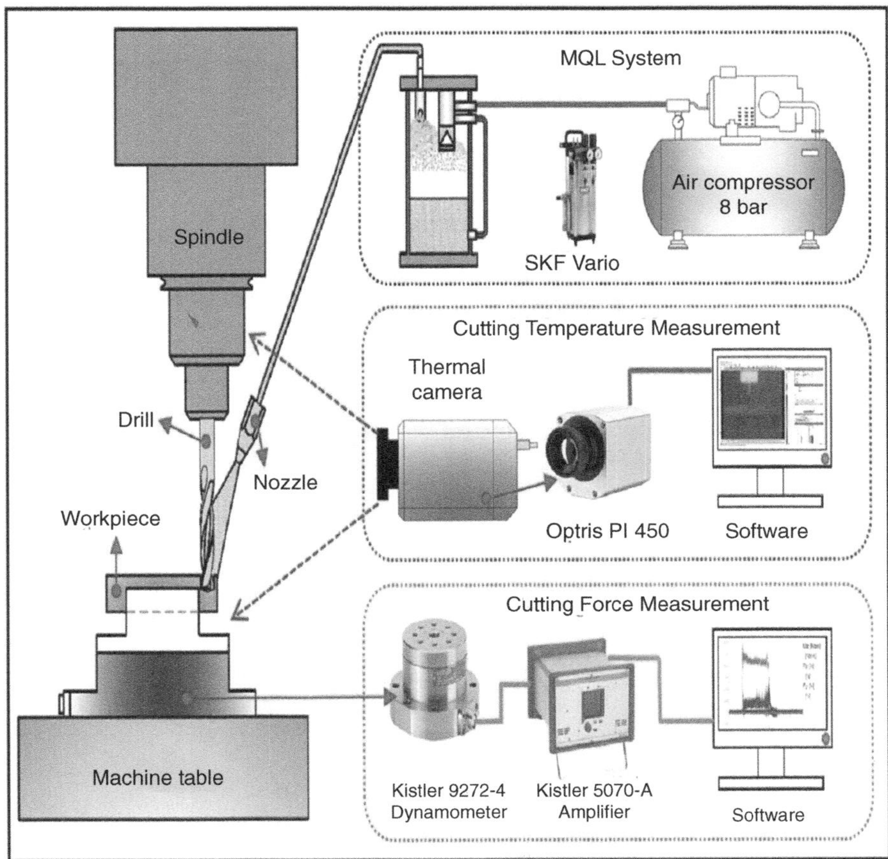

FIGURE 7.5 Application of nanofluid Machining (Şirin et al., 2021)

Nanofluid mixes are created by mixing solid nanoparticles with cutting fluid to improve the cutting oil quality. Nanoparticles improve the thermal conductivity and tribological properties of fluids. The heat transmission rate in the cutting region will also increase if the cutting fluids have a high thermal conductivity. Also, a more effective oil film layer forms between the tool-chip and the tool-material interfaces due to the enhancement of the tribological characteristics of cutting oils (Das, 2017; Sharma et al., 2015b). As a result, the unequal transmission of cutting temperature to the surfaces of the cutting tool and material may be avoided. Cutting force, surface quality, and cutting tool life can all result in significant gains (Hwang et al., 2006).

It is preferable to prepare nanofluids in either one step or two steps. Nanoparticles are produced in one process and dispersed throughout the base fluid simultaneously. Preparing metal nanoparticles in a single stage is advised to avoid oxidation impact and ensure excellent thermal conductivity. Nevertheless, the one-step

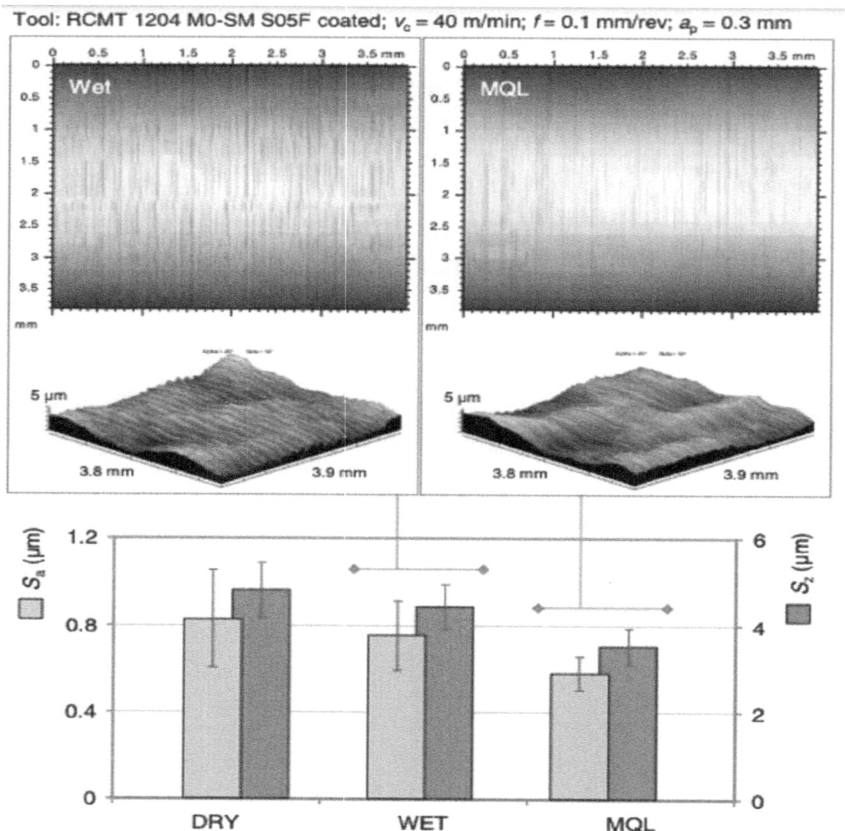

FIGURE 7.6   Roughness values turned in different cooling conditions (Priarone et al., 2014)

approach is not favored since it delays manufacturing, necessitates costly preparation procedures, and needs a vacuum to make nanofluids (Das, 2017). Nanofluids are prepared using the two-step process in two phases. This process involves incorporating the fully prepared solid nanoparticles into the base liquid at different velocities. In the two-step method, other mixing techniques, such as magnetic stirrer, ultrasonic homogenizer, ultrasonic bath, etc., are used to obtain nanofluids (Devendiran and Amirtham, 2016).

Some chemicals (i.e., surface agents-surfactants) can be doped at the nanofluid preparation stage because surfactants help to create a homogeneous mixture and reduce the surface tension. Surfactants suspend the nanoparticles, ensuring a homogeneous mixture, preventing agglomeration, and forming consistent combinations for a relatively long term (Jiang et al., 2003; Tanvir and Qiao, 2012).

7.2.3.1   NanoMQL Machining Case Study 1

Inconel X-750 material was milled by Şirin and Kıvak under dry, MQL, and nanofluid conditions, adding hBN, MoS$_2$, and graphite nanoparticles. Nanoparticles were

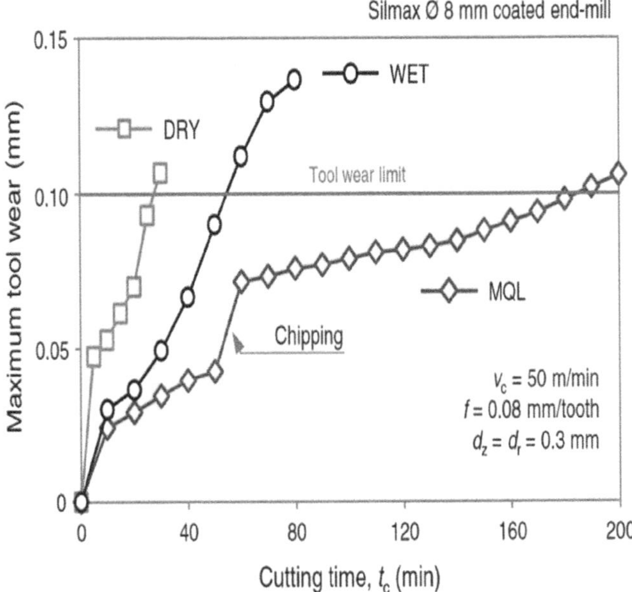

FIGURE 7.7 Maximum flank wear values when milling under different cooling conditions (Priarone et al., 2014)

doped to the base oil at several ratios (0.25, 0.5, 0.75, and 1 vol%). Performance metrics for surface roughness, cutting temperature, cutting force, and tool life were developed. According to the authors, nanofluid cooling conditions are superior to dry and MQL cooling conditions in terms of efficiency. Researchers also asserted that the hBN-enhanced nanofluid cooling condition was superior to other cooling conditions among the nanofluid cooling conditions (Figures 7.8 and 7.9) (Şirin and Kıvak, 2019).

7.2.3.2 Hybrid NanoMQL Machining

Nanoparticles can have different superior properties, such as thermal conductivity, lubricity, adhesion, corrosion prevention, etc., relative to each other. Nanoparticles can have different forms, properties, and capabilities for transfer energy. Therefore, the use of hybrid nanofluids, which join outstanding characteristics of each one, has increased recently. Hybrid nanofluids are obtained by adding at least two nanoparticles with different properties to the base liquid. More effective nanofluids are obtained by combining different superior properties compared to each other. The hybrid nanofluid preparation process is given in Figure 7.10.

7.2.3.3 Hybrid NanoMQL Machining Case Study 1

Hastelloy X superalloy material drilled under dry, MQL, and Nano MQL (hBN, GNP, and hBN/GNP) cooling conditions by Şirin et al. The factors under investigation

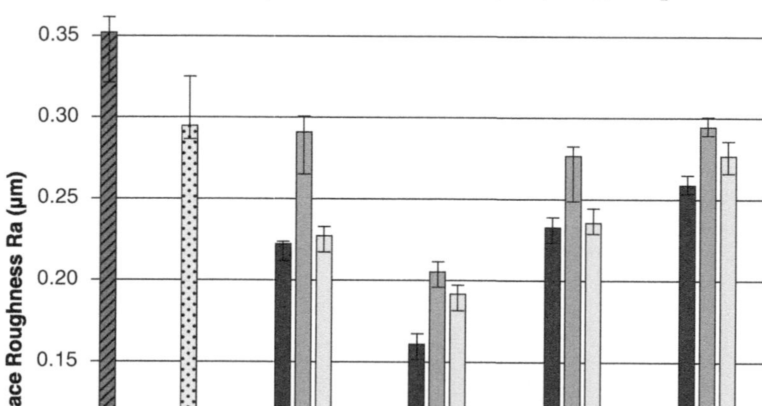

FIGURE 7.8 Surface roughness rates vary depending on the cutting situations (Şirin and Kıvak, 2019)

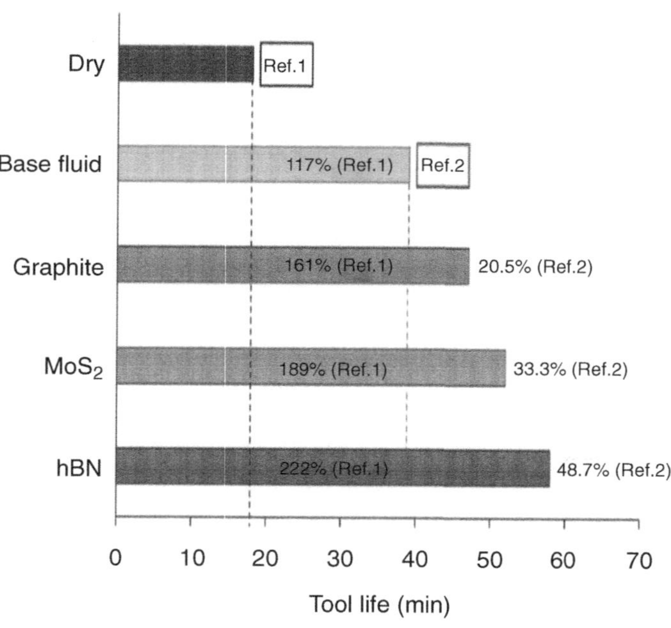

FIGURE 7.9 Tool life values depend on cutting conditions (Şirin and Kıvak, 2019)

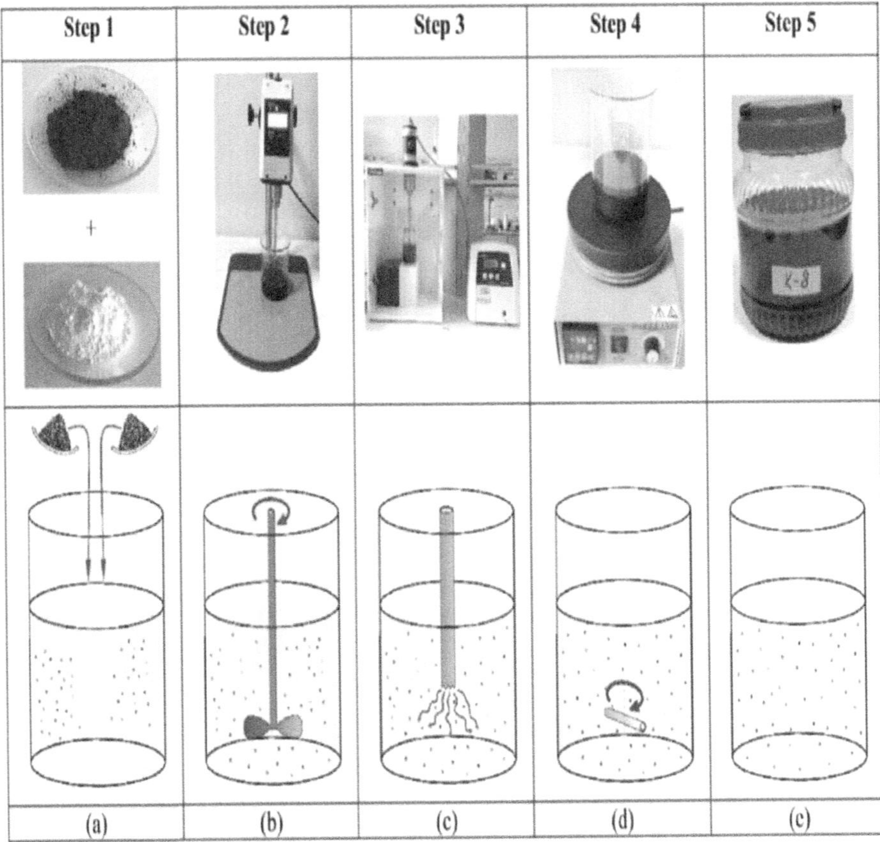

FIGURE 7.10   Hybrid nanofluid preparation steps, (a) Adding nanoparticles (b) Mixing with homogenizer (c) Ultrasonic mixing (d) Magnetic mixing (e) Prepared nanofluids (Şirin and Kıvak, 2021)

in this study were cutting force, hole quality, cutting temperature, burr height, tool wear, and tool life. Based on the experimental findings, the researchers claimed that the hybrid NanoMQL (hBN/GNP) condition increased tool life and significantly improved surface roughness and hole quality values (Figures 7.11 and 7.12) (Şirin et al., 2021).

7.2.4   CRYOGENIC MACHINING

The cutting tool and workpiece are cooled in metalworking operations using cryogenic cooling, an environmentally friendly cooling technique. As cryogens for chilling, gases like nitrogen (N_2), helium (He), carbon dioxide (CO_2), and argon (Ar) are often favored (Şirin, 2022; Yildiz and Nalbant, 2008). Liquid nitrogen, LN_2 ($-196°C$), and CO_2 ($-80°C$) are the most widely used gases in cryogenic refrigeration because

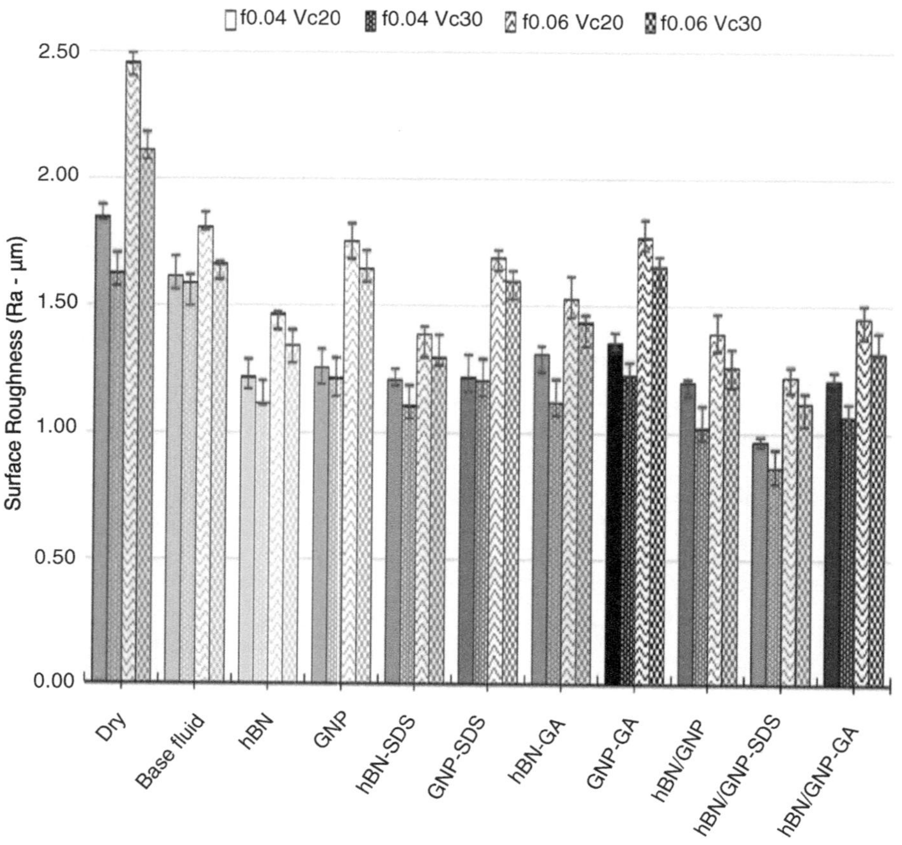

FIGURE 7.11 Surface roughness values according to different cutting conditions (Şirin et al., 2021)

they are colorless, odorless, tasteless, and non-toxic properties. Seventy-eight percent of the atmosphere is nitrogen (Gupta et al., 2021; Hong, 2001). Furthermore, LN_2 and CO_2 swiftly vaporize and vanish into the environment, leaving no residue to contaminate the material, chips, machine, operator, or operator's workspace. So, cryogenic cooling eliminates disposal costs (Umbrello et al., 2012). Cryogenic machining may also be utilized to process work materials at faster cutting rates, achieving greater surface integrity and higher surface quality, increasing machinability, and lowering total cost (Huang et al., 2022).

The heat produced between the tool-workpiece and the chip is reduced via cryogenic cooling. The use of cryogenic cooling during machining can save tool wear, avoid lubricant failure due to heat, and strengthen chip cutting fracture. Cryogens include gases like carbon dioxide (CO_2) and liquid nitrogen (LN_2). LN_2 and CO_2

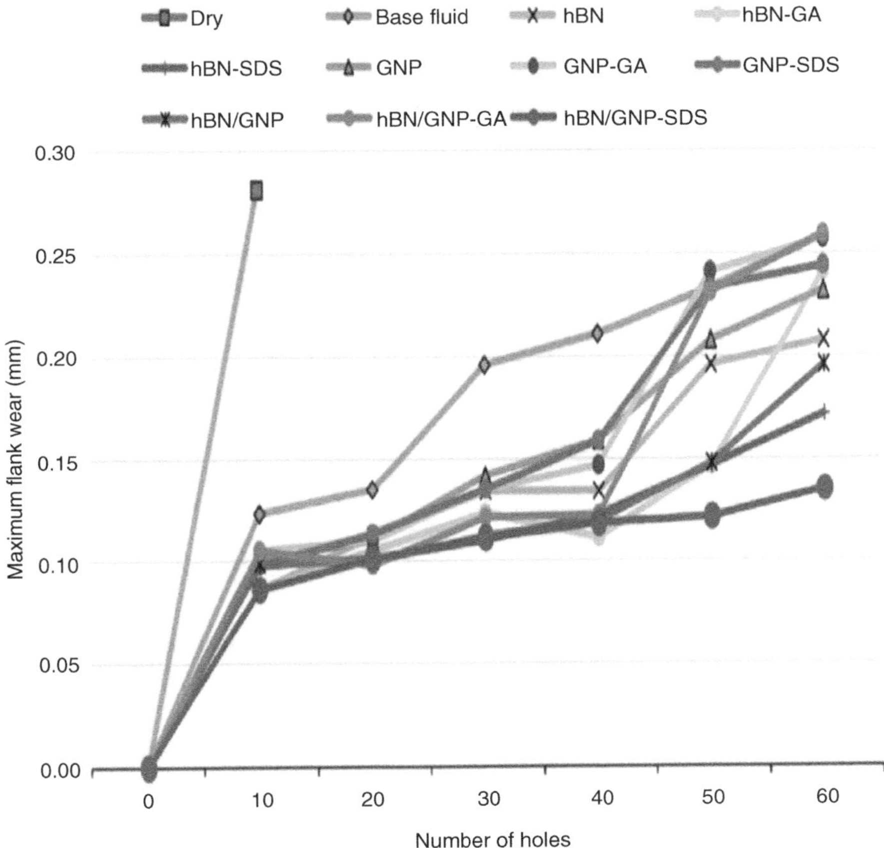

FIGURE 7.12 Maximum flank wear values according to different cutting conditions (Şirin et al., 2021)

offer to use at different temperatures. CO_2 and LN_2 phase diagrams are given in Figure 7.13.

The low-temperature generation mechanisms of LN_2 and CO_2 cryogens used as refrigerants are different from each other. These differences are observed in the phase diagrams. CO_2 is sent to the cutting area as a liquid from medium-pressure tanks at a pressure of about 57 bar and a temperature of 20°C. As the liquid CO_2 comes out of the cooling channels, it enlarges, and the pressure drops to ambient pressure ($P = 1.013$ bar). The environment cools to −78.5°C due to the Joule Thomson effect, and CO_2 turns from solid to gas. The higher ambient temperature of the cutting area causes the solid phase to sublimate without leaving any residue, resulting in lower process temperatures. LN_2 is stocked in isolated tanks. It transforms from solid to liquid phase at ambient pressure at 210°C and starts to

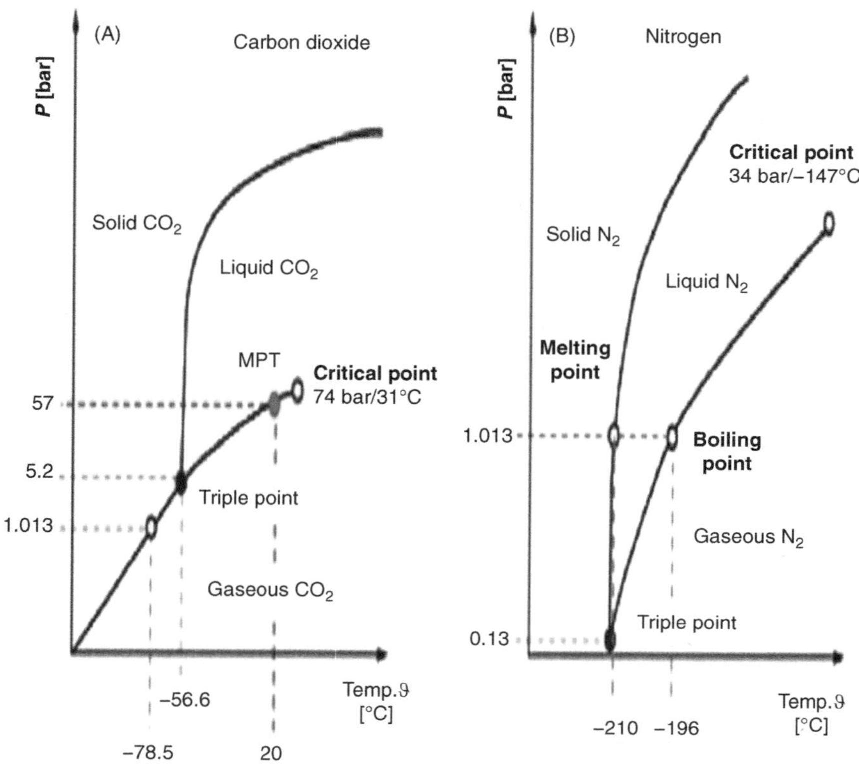

FIGURE 7.13 Phase diagram for (a) CO_2 and (b) LN_2 (Blau et al., 2015)

boil at −196°C. It can be said that LN_2 is a better cooling condition compared to CO_2 because of low temperatures (Bayraktar, 2000). Low LN_2 temperatures under atmospheric pressure can be problematic in its practice as a refrigerant. Supply valves, pipes, hoses, cutting tools, and machine parts must be preserved from potential hazards and the impact of low temperatures for cooling efficiency (Blau et al., 2015). For this purpose, vacuum jacket hoses can minimize thermal losses (Yıldırım, 2019).

The tested benefits of cryogenic machining, in general, contain:

- Enhanced process sustainability (ecological processes with no adverse health effects for employees, cleaner and safer),
- Enhanced material removal rate (MRR) with no enhancement in tool wear ratios and tool change period, decreasing total costs through higher efficiency,
- Longer cutting tool life due to less wear and chemical wear,
- Enhanced product quality on machined parts thanks to the elimination of mechanical and chemical deterioration,

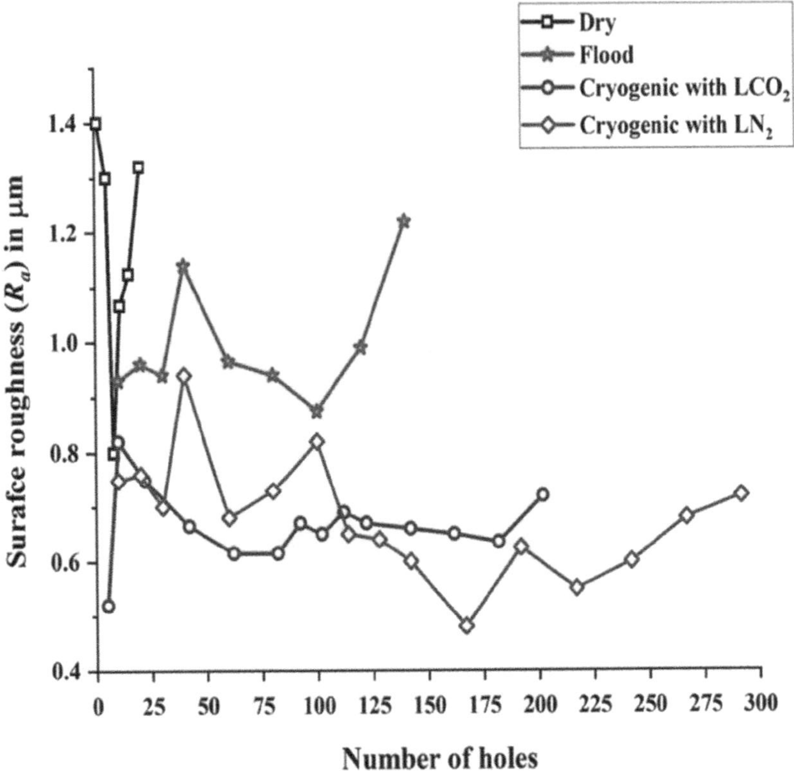

FIGURE 7.14   Surface roughness values according to various cutting conditions (Shah and Khanna, 2020)

- Phase changes with improved surface integrity, more suitable dynamic recrystallization, corrosion, and wear resistance through controllable microstructure (Umbrello et al., 2012).

7.2.4.1   Cryogenic Machining Case Study 1

Shah et al. drilled the VT-20 alloy material under dry, flood, and cryogenic (LN$_2$, CO$_2$) cooling conditions. The experimental study chose surface finish, hole quality, microhardness, chip, and power consumption analysis as performance criteria. Dry and flood cooling conditions were compared with cryogenic cooling conditions. The authors claimed that due to the experimental study, cryogenic cooling condition, a green cooling technique, exhibits very efficient results on account of tool wear and power consumption (Figures 7.14 and 7.15) (Shah and Khanna, 2020).

7.2.5   HYBRID COOLING/LUBRICATION MACHINING

The hybrid cooling method is a superior cooling method obtained by combining the capabilities of the MQL and/or nanofluid (mono+hybrid) method with the features

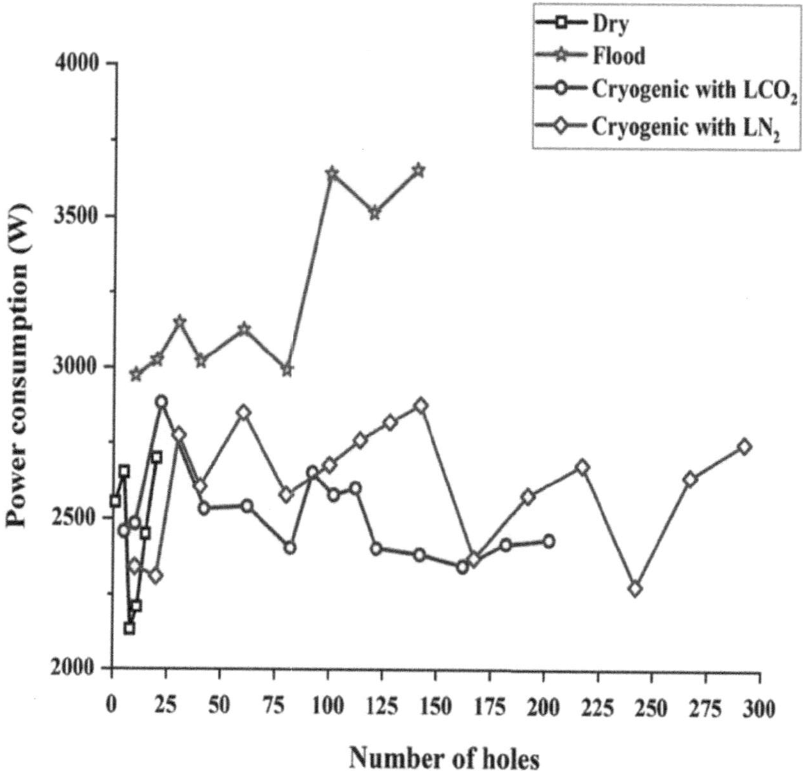

FIGURE 7.15 Power consumption values according to various cutting conditions (Shah and Khanna, 2020)

of the cryogenic cooling method. MQL+cryogenic, nanofluid+cryogenic, and hybrid nanofluid+cryogenic techniques are examples of hybrid cooling/lubrication techniques (Figure 7.16).

In the hybrid cooling/lubrication method, lubrication is provided by MQL and nanofluids, and cryogenic gases provide cooling. Hybrid methods, in which lubrication and cooling are provided simultaneously, show superior performance compared to other methods.

7.2.5.1 Hybrid Cooling/Lubrication Machining Case Study 1

Şirin turned Haynes 25 superalloy material under dry, MQL and NanoMQL (GnP, MWCNT, GnP/MWCNT), cryogenic (N₂), and hybrid cooling conditions. The author examined the surface roughness, surface topography, cutting temperature, tool wear, and chip form morphology in the experimental study. When the test results were evaluated, the researcher asserted that the hybrid (MWCNT/GnP+N2) condition reduced the Ra value by 36.36% and the worn tool values by 45.13% (Figures 7.17 and 7.18) (Şirin, 2022).

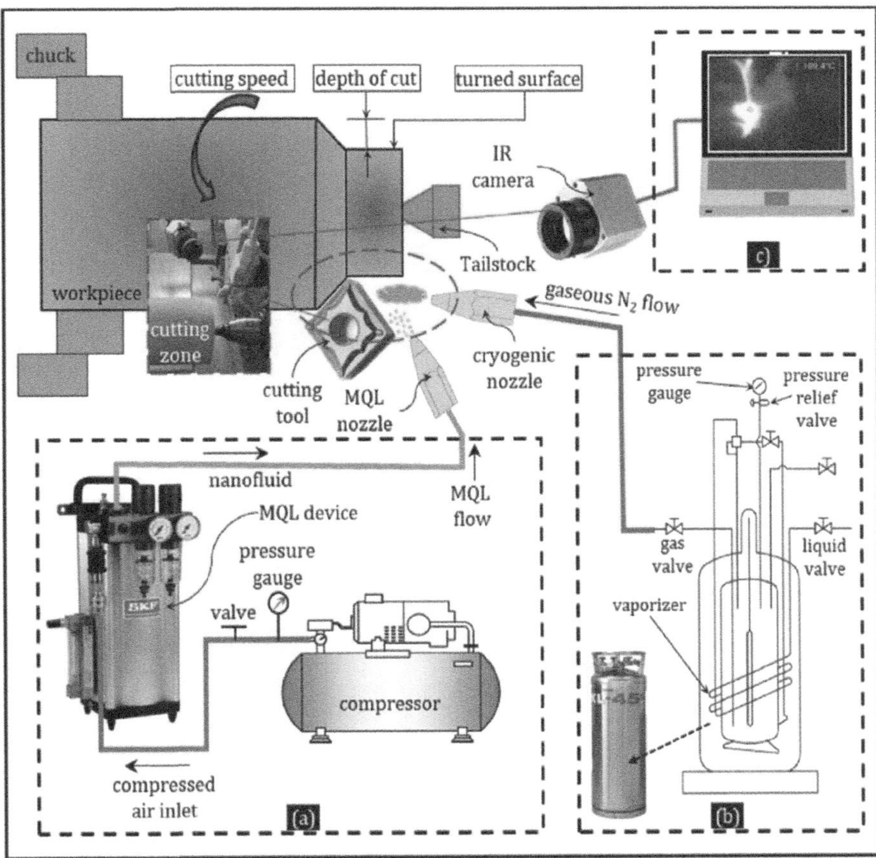

FIGURE 7.16 Application of hybrid cooling/lubrication machining (MQL machining+ cryogenic machining (Şirin, 2022))

7.3 GENERAL COMPRESSION OF SUSTAINABLE MACHINING

The Green Machining method has grown in popularity in recent years. The cooling/lubrication processes used in Green Machining (Where no or little cutting fluid is used) are acceptable for the environment, health of the employees, and machining costs. These techniques send the coolants directly to the cutting area, so Green Machining techniques provide more effective cooling.

The MQL method is preferred over a conventional wet cutting condition (Generally called as flood cooling) because of human health and the environment and uses very little cutting oil. However, nanoparticles are doped to the cutting oil to enhance the performance and the tribological effects of the cutting oil in heavy machining conditions. In this way, the tribological and lubrication performance of nanofluids is increased. Cryogenic cooling ensures sufficient advancements in critical aspects such as less cutting temperature, improving surface quality, enhancing

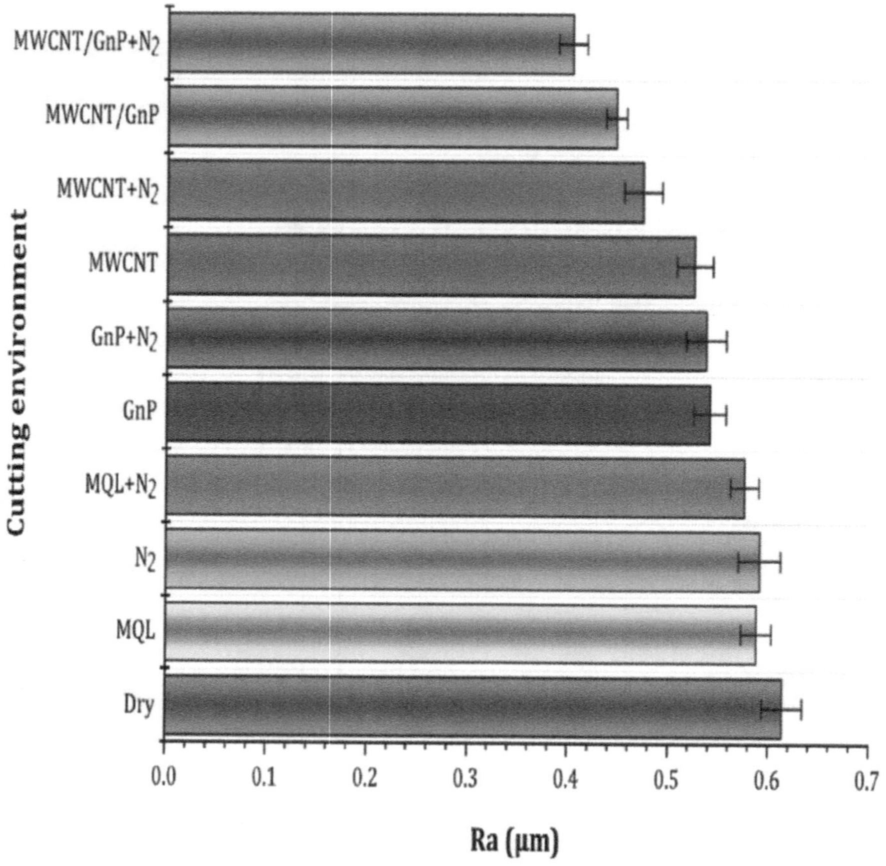

FIGURE 7.17   Surface roughness values according to several cutting conditions (Şirin, 2022)

tool life, and decreasing tool wear. Moreover, cryogenic cooling is considered a Green Machining technology and a critical cooling method for sustainable production (Nalbant and Yildiz, 2011). Hybrid cooling/lubrication is a highly effective Green Machining method that provides both lubrication and cooling for machining efficiency (Table 7.1).

7.4 CONCLUSION

This book chapter provides an overview of sustainable or Green Machining technologies strategies. Information on Dry, MQL, NanoMQL, Cryogenic, and Hybrid cooling/lubrication strategies is provided and supported by case studies. The general results obtained are given below.

- Dry machining offers an ecological environment but is inadequate in terms of machining efficiency.

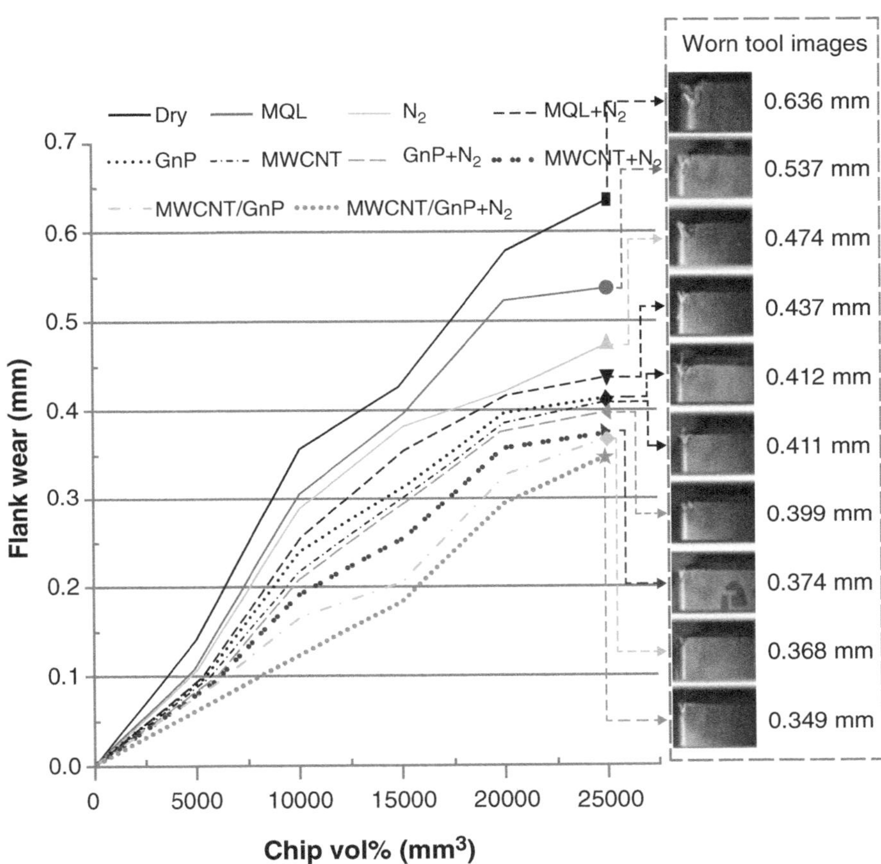

FIGURE 7.18 Flank wear results under different cutting environments (Şirin, 2022)

- MQL machining with 50/50 dry/oil conditions has come to the fore in sustainable manufacturing, but MQL machining has lagged behind flood lubrication.
- NanoMQL provided efficient machining with mechanisms: mending, polishing, rolling, cushioning, interlayer sliding, etc. NanoMQL machining has delivered significant improvements by affecting the cutting tool-workpiece interfaces. However, under heavy machining conditions, the cooling performance is relatively low.
- Cryogenic machining is among the most effective cooling strategies for reducing the temperature in the cutting zone. However, since there is no lubricating agent, the wear mechanisms in the cutting tool grow rapidly.
- Hybrid cooling/lubrication strategies are an approach that combines MQL/NanoMQL and cryogenic machining strategies. This approach performs both cooling and lubrication simultaneously in the cutting zone. In this way, an efficient process occurs under heavy machining conditions.

TABLE 7.1

Application and efficiency of various cooling/lubrication machining (Jawahir et al., 2016)

	Effect of the cooling and lubricating strategy	Flood (emulsion/ oil)	Dry (compressed air)	MQL (oil)	Cryogenic (LN$_2$)	Hybrid (LN$_2$ + MQL)
Primary	Cooling	Good	Poor	Marginal	Excellent	Excellent
	Lubrication	Excellent	Poor	Excellent	Marginal	Excellent
	Chip removal	Good	Good	Marginal	Good	Good
	Machine cooling	Good	Poor	Poor	Marginal	Marginal
Secondary	Workpiece cooling	Good	Poor	Poor	Good	Good
	Dust/particle control	Good	Poor	Marginal	Marginal	Good
	Product quality (surface integrity)	Good	Poor	Marginal	Excellent	Excellent
	Sustainability concerns	Water pollution, microbial infestation, and high cost	Poor surface integrity due to thermal damage	Harmful oil vapor	Initial cost	Initial cost, oil vapor

The evaluations showed that sustainable manufacturing technologies would soon show significant developments in the aerospace/aviation, defense, and heavy processing industries. Future studies should focus on different hybrids (high-pressure MQL/NanoMQL, extra Cryogenic cooling, etc.) cooling/lubrication strategies.

REFERENCES

Anandan, V., Babu, M., Sezhian, M., Yildirim, C. V., & Dinesh Babu, M. (2021). Influence of graphene nanofluid on various environmental factors during turning of M42 steel. *Elsevier Journal of Manufacturing Processes*, *68*, 90–103.

Bayraktar, Ş. (2000). Cryogenic cooling-based sustainable machining.In *High Speed Machining*, edited by Kapil Gupta and Joao Paulo Davim, 223–241. Elsevier Academic Press.

Blau, P., Busch, K., Dix, M., Hochmuth, C., Stoll, A., & Wertheim, R. (2015). Flushing strategies for high performance, efficient and environmentally friendly cutting. *Elsevier, Procedia CIRP*, *26*, 361–366.

Das, P. (2017). A review based on the effect and mechanism of thermal conductivity of normal nanofluids and hybrid nanofluids. *Journal of Molecular Liquids*, *240*, 420–446.

Devendiran, D. K., & Amirtham, V. A. (2016). A review on preparation, characterization, properties and applications of nanofluids. In *Renewable and Sustainable Energy Reviews* (Vol. 60, pp. 21–40). Elsevier Ltd. https://doi.org/10.1016/j.rser.2016.01.055

Devillez, A., Le Coz, G., Dominiak, S., & Dudzinski, D. (2011). Dry machining of Inconel 718, workpiece surface integrity. *Elsevier Journal of Materials Processing Technology*, *211*(10), 1590–1598.

Fratila, D., & Production, C. C. (2011). Application of Taguchi method to selection of optimal lubrication and cutting conditions in face milling of AlMg$_3$. *Elsevier, Journal of Cleaner Production*, *19*(6–7), 640–645.

Ghadimi, A., Saidur, R., & Metselaar, H. S. C. (2011). A review of nanofluid stability properties and characterization in stationary conditions. *International Journal of Heat and Mass Transfer*, *54*(17–18), 4051–4068. https://doi.org/10.1016/J.IJHEATMASSTRANSFER.2011.04.014

Gupta, K. (2020). A review on Green machining techniques. *Procedia Manufacturing*, *51*, 1730–1736. https://doi.org/10.1016/J.PROMFG.2020.10.241

Gupta, M. K., Song, Q., Liu, Z., Sarikaya, M., Mia, M., Jamil, M., Singla, A. K., Bansal, A., Pimenov, D. Y., & Kuntoğlu, M. (2021). Tribological performance based machinability investigations in cryogenic cooling assisted turning of α-β titanium Alloy. *Tribology International*, *160*, 107032. https://doi.org/10.1016/J.TRIBOINT.2021.107032

Haider, J., & Hashmi, M. S. J. (2014). 8.02—Health and environmental impacts in metal machining processes. In *Comprehensive Materials Processing*, edited by Saleem Hashmi, Gilmar Ferreira (pp. 7–33). Elsevier.

Hong, S. Y. (2001). Economical and ecological cryogenic machining. *The Journal of Manufacturing Science and Engineering*, *123*(2), 331–338.

Huang, H., Liu, S., Zhu, L., Qing, Z., Bao, H., & Liu, Z. (2022). Cooling and lubrication performance of supercritical CO$_2$ mixed with nanofluid minimum quantity lubrication in turning Ti-6Al-4V. *International Journal of Advanced Manufacturing Technology*, *122*(7–8), 2927–2938. https://doi.org/10.1007/S00170-022-10091-9

Hwang, Y., Park, H. S., Lee, J. K., & Jung, W. H. (2006). Thermal conductivity and lubrication characteristics of nanofluids. *Current Applied Physics*, *6*, e67–e71. https://doi.org/10.1016/j.cap.2006.01.014

Jain, A., & Kansal, H. (2017). Green machining–Machining of the future. *In 4th National Conference on Advancements in Simulation & Experimental Techniques in Mechanical Engineering (NCASEme-2017)*, 21–25. Chandigarh University.

Jawahir, I. S., Attia, H., Biermann, D., Duflou, J., Klocke, F., Meyer, D., Newman, S. T., Pusavec, F., Putz, M., Rech, J., Schulze, V., & Umbrello, D. (2016). Cryogenic manufacturing processes. *CIRP Annals – Manufacturing Technology*, *65*(2), 713–736. https://doi.org/10.1016/j.cirp.2016.06.007

Jiang, L., Gao, L., & Sun, J. (2003). Production of aqueous colloidal dispersions of carbon nanotubes. *Journal of Colloid and Interface Science*, *260*(1), 89–94. https://doi.org/10.1016/S0021-9797(02)00176-5

Mia, M., Khan, M. A., & Dhar, N. R. (2017). Study of surface roughness and cutting forces using ANN, RSM, and ANOVA in turning of Ti-6Al-4V under cryogenic jets applied at flank and rake faces of coated WC tool. *The International Journal of Advanced Manufacturing Technology*, *93*(1–4), 975–991. https://doi.org/10.1007/s00170-017-0566-9

Nalbant, M., & Yildiz, Y. (2011). Effect of cryogenic cooling in milling process of AISI 304 stainless steel. *Elsevier Transactions of Nonferrous Metals Society of China*, *21*, 72–79. https://doi.org/10.1016/S1003-6326(11)60680-8

Priarone, P., Robiglio, M., Settineri, L., & Tebaldo, V. (2014). Milling and turning of titanium aluminides by using minimum quantity lubrication. *Elsevier Procedia CIRP*, *24*, 62–67.

Pusavec, F., Kramar, D., Krajnik, P., & Kopac, J. (2010). Transitioning to sustainable production – Part II: Evaluation of sustainable machining technologies. *Journal of Cleaner Production*, *18*(12), 1211–1221. https://doi.org/10.1016/J.JCLEPRO.2010.01.015

Sartori, S., Ghiotti, A., & Bruschi, S. (2018). Solid lubricant-assisted minimum quantity lubrication and cooling strategies to improve Ti6Al4V machinability in finishing turning. *Tribology International, 118*, 287–294. https://doi.org/10.1016/J.TRIBOINT.2017.10.010

Shah, P., & Khanna, N. (2020). Comprehensive machining analysis to establish cryogenic LN_2 and LCO_2 as sustainable cooling and lubrication techniques. *Tribology International, 148*, 1–15.

Sharma, A. K., Tiwari, A. K., & Dixit, A. R. (2015a). Mechanism of nanoparticles functioning and effects in machining processes: A review. *Undefined, 2*(4–5), 3539–3544. https://doi.org/10.1016/J.MATPR.2015.07.331

Sharma, A. K., Tiwari, A. K., & Dixit, A. R. (2015b). Progress of nanofluid application in machining: A review. *Materials and Manufacturing Processes, 30*(7), 813–828. https://doi.org/10.1080/10426914.2014.973583Şirin, E., Kıvak, T., & Yıldırım, Ç. V. (2021). Effects of mono/hybrid nanofluid strategies and surfactants on machining performance in the drilling of Hastelloy X. *Tribology International, 157*, 106894. https://doi.org/10.1016/J.TRIBOINT.2021.106894

Şirin, Ş. (2022). Investigation of the performance of cermet tools in the turning of Haynes 25 superalloy under gaseous N2 and hybrid nanofluid cutting environments. *Journal of Manufacturing Processes, 76*, 428–443. https://doi.org/10.1016/J.JMAPRO.2022.02.029

Şirin, Ş., & Kıvak, T. (2019). Performances of different eco-friendly nanofluid lubricants in the milling of Inconel X-750 superalloy. *Tribology International, 137*, 180–192. https://doi.org/10.1016/j.triboint.2019.04.042

Şirin, Ş., & Kıvak, T. (2021). Effects of hybrid nanofluids on machining performance in MQL-milling of Inconel X-750 superalloy. *Journal of Manufacturing Processes, 70*, 163–176. https://doi.org/10.1016/J.JMAPRO.2021.08.038

Sreejith, P. S., & Ngoi, B. K. A. (2000). Dry machining: Machining of the future. *Elsevier Journal of Materials Processing Technology, 101*(1–3), 287–291.

Tai, B., Stephenson, D., Furness, R., & Cirp, A. S. (2014). Minimum quantity lubrication (MQL) in automotive powertrain machining. *Elsevier Procedia Cirp, 4*, 523–528.

Tanvir, S., & Qiao, L. (2012). Surface tension of nanofluid-type fuels containing suspended nanomaterials. *Nanoscale Research Letters, 7*(1), 226. https://doi.org/10.1186/1556-276X-7-226

Umbrello, D., Micari, F., & Jawahir, I. S. (2012). The effects of cryogenic cooling on surface integrity in hard machining: A comparison with dry machining. *CIRP Annals, 61*(1), 103–106. https://doi.org/10.1016/J.CIRP.2012.03.052

Xie, H., Jiang, B., He, J., Xia, X., & Pan, F. (2016). Lubrication performance of MoS_2 and SiO_2 nanoparticles as lubricant additives in magnesium alloy-steel contacts. *Tribology International, 93*, 63–70. https://doi.org/10.1016/J.TRIBOINT.2015.08.009Yıldırım, Ç. V. (2019). Experimental comparison of the performance of nanofluids, cryogenic and hybrid cooling in turning of Inconel 625. *Tribology International, 137*, 366–378. https://doi.org/10.1016/j.triboint.2019.05.014

Yıldırım, Ç. V., Sarıkaya, M., Kıvak, T., & Şirin, Ş. (2019). The effect of addition of hBN nanoparticles to nanofluid-MQL on tool wear patterns, tool life, roughness and temperature in turning of Ni-based Inconel 625. *Tribology International, 134*, 443–456. https://doi.org/10.1016/j.triboint.2019.02.027

Yildiz, Y., & Nalbant, M. (2008). A review of cryogenic cooling in machining processes. *International Journal of Machine Tools and Manufacture, 48*(9), 947–964.

8 Environmentally Benign Machining of Composites

P. M. Gopal, V. Kavimani and S. Sudhagar

8.1 INTRODUCTION

The concept of "sustainability" has been widely discussed and studied through various means like conferences and debates. While different experts have tried to explain and define the term "sustainability" in their individual ways, Dr. Brundtland's definition from the report "Our Common Future" is the most frequently cited [1]. This definition emphasizes that growth should not come at the expense of the well-being of future generations [2]. The environmentally benign manufacturing and sustainable manufacturing are two very comparable ideas that are frequently used interchangeably, yet they differ significantly in key ways [3]. In general, it can be stated that the environmentally benign manufacturing is a part of sustainable manufacturing. Sustainable manufacturing refers to manufacturing processes that are created to satisfy current needs without jeopardizing the potential of future generations to satisfy their own needs [4]. Sustainable manufacturing takes a more holistic view of the manufacturing process, considering not only the ecological impact, but also the societal and financial impacts of the process. Sustainable manufacturing seeks to optimize the use of resources, reduce waste and emissions, and ensure that the manufacturing process is socially responsible and economically viable in the long term [5].

Environmentally benign manufacturing, on the other hand, refers to manufacturing processes that are designed to minimize the environmental impact of the manufacturing process. This may include reducing the use of hazardous materials, minimizing waste generation, and reducing energy consumption [6]. The goal of environmentally benign manufacturing is to reduce the environmental footprint of the manufacturing process, without necessarily considering the long-term sustainability of the process. While sustainable manufacturing takes a more comprehensive approach, considering the long-standing sustainability of the process from economic, communal, and environmental perception, environmentally benign manufacturing focuses on minimizing the environmental impact of the manufacturing process.

Manufacturing is a major contributor to energy and natural resource consumption, and sustainability in this sector is critical. Manufacturing is energy-intensive and has a significant environmental impact [7]. In every country, manufacturing accounts major part of their GDP and concurrently more total energy usage, but it also contributes to more greenhouse gas emissions and has a significant environmental impact due to waste production. As a result, finding sustainable and greener ways to manufacture is a continual focus of the scientific community.

DOI: 10.1201/9781003352402-8

Conventional machining is the process of removing material layers using a pointed, wedge-shaped tool via plastic deformation and consequent shearing. To cool the tool and work-piece during this process, a liquid coolant, known as cutting fluid, is traditionally used to take away the heat generated throughout the metal removal process [8, 9]. Cutting fluids, also referred to as flood coolants, are continuously applied in large volumes (In litres per minute) at the machining location of the work piece [10]. In addition to cooling, cutting fluids reduce area of contact and friction between the work and tool, further lowering heat and tool temperature. Cutting fluid also effectively removes chips and swarf. The temperature of tool, life of tool, surface roughness of the product, residual stresses, and accuracy of the finished part, all affect the productivity of the shop floor, and the cutting fluid has a significant impact on these parameters [11]. The lubricating oil and additives in the cutting fluid form a shield that protects the workpiece, cutting tool, and machining machinery against rust and corrosion. Every machine has a specific coolant handling system that allows cutting fluids to be reused several times, but doing so imposes a substantial and growing economic and environmental burden. More than 2 billion litres of plain cutting oils and water-based emulsions are consumed globally each year, and the equipment needed to use these fluids is expensive and energy-intensive. Cutting fluid formulations must be discarded after a specific amount of use since they can get contaminated with metal particles, ions, and bacterial and fungal development. To avoid pollution and ecological harm, large manufacturing enterprises must safely dispose of thousands of litres of waste cutting fluids. Up to 17% of the overall cost of the machined product can be attributed to the cost of cutting fluids and the systems used to regulate them. Cutting fluids can also result in risky working circumstances for employees owing to skin contact with microbes and biocides as well as inhalation of harmful metal particles, which can result in occupational illnesses. Strict environmental rules by statutory bodies have made research required in order to decrease the usage of dangerous cutting fluids in machining operations [12].

Aside from the traditional methods of machining composites, electric discharge machining (EDM) is a commonly used method due to its ability to machine hard materials that are difficult to work with using conventional methods, as well as its capability to produce complex shapes with both micro and macro removal rates, and work with different types of materials. There have been numerous successes in producing both macro and micro parts for industrial applications using EDM [13, 14]. The process of EDM involves the removal and shaping of hard material through repeated electrical discharges (Sparks) between an electrode (Tool) and the anode (Workpiece) in the presence of a dielectric fluid. As the dielectric fluid ionizes, heat and pressure increase in the gap between the anode and the electrode, which creates a path for each discharge. Despite its effectiveness, EDM has a significant environmental impact. Toxic emissions are produced during the machining process due to the temperature generated, and heavy, poorly biodegradable materials are mixed in the slurry. Additionally, electromagnetic radiation is emitted, which can harm the environment. High consumption of electricity is also a concern [15].

To address these environmental concerns, researchers have explored methods of machining components without cutting fluids, with minimal fluids, or with bio oils as a cutting fluid or dielectric medium. Machining without the use of cutting fluids is known as "dry machining," while machining with minimal cutting fluids is called "Minimum Quantity Lubrication" (MQL) machining. This chapter examines the use of environmentally friendly dry, MQL, and bio-oil-based machining techniques for composites.

8.2 COMPOSITES AND THEIR MACHINING

8.2.1 Composite Materials

Composite materials are made up of two or more separate materials, each having unique physical or chemical properties that are combined to produce a new material with better qualities than the sum of its parts [16]. The individual materials that make up the composite are called constituents, and they can be in various forms, such as fibres, particles, or flakes. Composite materials are designed to enhance specific properties of each constituent material while minimizing their weaknesses. For example, the strength and stiffness of fibres can be combined with the toughness of a polymer matrix to create a composite material that is stronger and more durable than either material alone. Other benefits of composite materials include corrosion resistance, lightweight, and fatigue resistance [17].

Composite materials are used in extensive array of applications, from aerospace and automotive industries to construction and sports equipment. Composite materials are highly versatile and offer a range of advantages over traditional materials, making them an important area of research and development in many industries [18]. Based on the kind of matrix employed, they may be divided into three categories: polymer matrix composites, metal matrix composites, and ceramic matrix composites. The choice of matrix material depends on the specific application and the desired properties of the composite material [19]. There are several types of composite materials, including:

Polymer Matrix Composites (PMCs): These composites are made by combining a polymer resin, such as epoxy or polyester, with a reinforcing material, such as fibreglass or carbon fibre. PMCs are lightweight, strong, and have good fatigue resistance, making them useful in applications such as aerospace, automotive, and sports equipment.

Metal Matrix Composites (MMCs): These composites are made by combining a metal matrix, such as aluminium, with a reinforcing material, such as SiC or TiC. MMCs have high strength, stiffness, and good thermal conductivity, making them useful in applications such as automotive and aerospace.

Ceramic Matrix Composites (CMCs): These composites are made by combining a ceramic matrix, such as silicon carbide or aluminium oxide, with a reinforcing material, such as carbon fibre or silicon carbide. CMCs have high strength, stiffness, and thermal stability, making them useful in applications such as gas turbine engines and rocket nozzles.

Carbon-Carbon Composites (C/C): These composites are made by combining carbon fibres with a carbon matrix. C/C composites have high strength, stiffness, and thermal stability, making them useful in applications such as high-temperature aerospace and defence applications.

Hybrid Composites: These composites are made by combining two or more diverse kinds of reinforcing materials, such as fibres and particles, with a matrix material. Hybrid composites can be tailored to specific applications and can have a range of properties, depending on the materials used.

8.2.2 MACHINING

Machining is important for any material because it allows the material to be shaped, formed, and finished to meet specific design requirements [20]. Machining involves removing material from a workpiece using various tools and techniques, such as cutting, drilling, milling, and grinding. Machining allows precise control over the size, shape, and surface finish of a material. This is important for ensuring that the material meets specific dimensional and tolerance requirements for its intended use. Machining can improve the surface finish of a material by removing roughness and other imperfections, resulting in a smoother and more uniform surface that can improve the material's performance and appearance [21]. Machining makes possibility for the formation of intricate profiles and features that may not be feasible with other fabrication processes. This is important for applications that require intricate and precise parts. Machining can also affect the material's properties, such as strength, hardness, and ductility. By selecting the appropriate machining method and conditions, it is possible to enhance or modify these properties to better suit the application. The followings are the mostly used machining processes for advanced materials.

Turning: To achieve a cylindrical form, turning is an operation that requires rotating a workpiece on a lathe and removing material with a single point cutting tool. The production of shafts, bolts, and other cylindrical components is frequently accomplished using this method.

Milling: Using a revolving cutting tool, the process of milling involves removing material from a workpiece. The forms that may be produced by this method include flat surfaces, slots, and curves. Milling machines are commonly used in manufacturing and metalworking.

Drilling: Drilling involves creating a hole in a workpiece using a rotating cutting tool. This process can be done by hand or with a drilling machine. Drilling is used to create holes for bolts, screws, and other fasteners.

Grinding: Usage of an abrasive wheel to remove material from a work piece is the process of grinding. This process is commonly used to create smooth surfaces and to sharpen cutting tools.

Electrical discharge machining (EDM): Electrical sparks are used in the machining process known as EDM to remove material from a work piece. This process is commonly used to produce complex shapes and to machine hard materials. It is employed to make complex profiles on difficult to machine materials like composites.

Laser Cutting: A strong laser beam is used in the process of "laser cutting" to cut through material. This process is commonly used to create precise cuts in metal, plastic, and other materials.

Water Jet Cutting: A high-pressure water jet is used in the process of water jet cutting to cut through material. This process is commonly used to cut through thick materials or materials that cannot be cut using traditional methods. It is chiefly used for machining polymer composites.

Here, machining of composite materials is important for several reasons. Composites can be made to meet specific requirements by varying the type of fibres, matrix material, and layup orientation. Machining allows for precise shaping and finishing of composite components to meet these requirements. Composite materials can be difficult to shape and finish due to their anisotropic nature and potential for delamination [22]. Machining techniques can help achieve precise dimensional accuracy while minimizing the risk of damage. Machining of composite materials allows for detailed inspection and quality control, ensuring that the finished components meet the required standards. Machining can be used to repair damaged composite components by removing the damaged area and replacing it with a new piece of composite material.

8.2.3 Challenges in Machining of Composites

Machining of composite materials is often more challenging than machining of traditional materials like metals [23], due to several reasons:

Anisotropic nature: Composite materials have anisotropic properties, which means they have different mechanical properties in different directions. This makes it difficult to achieve consistent cutting forces and tool wear during machining.

High abrasiveness: Composite materials are often made up of hard, abrasive fibres, such as carbon or glass fibres, which can cause excessive tool wear and result in poor surface finish.

Delamination: Delamination is a common problem during machining of composites. The layers of composite material can separate or break apart, leading to surface defects and weakening of the material. It causes the most substantial damage to a compromised composite. The layers of the composite laminate may separate. Delamination is brought on by matrix cracking, bending fractures, and shear cracks. It will progressively lower the laminate's compression strength, causing the composite to buckle. More severe damage results from the composite's response to stress, which starts fissures and results in delamination between the laminates. Figure 8.1 shows the delamination of laminated composite during drilling.

Heat generation: Composite materials are poor conductors of heat, so machining can generate excessive heat, leading to thermal damage and distortion of the part.

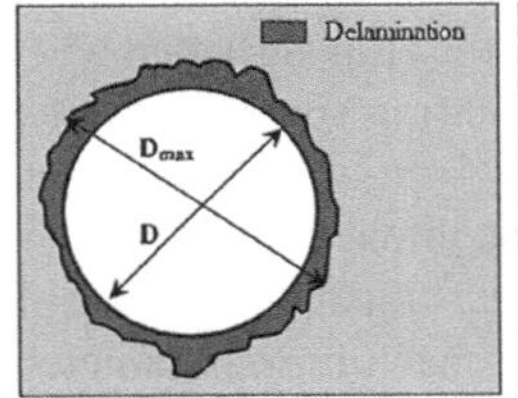

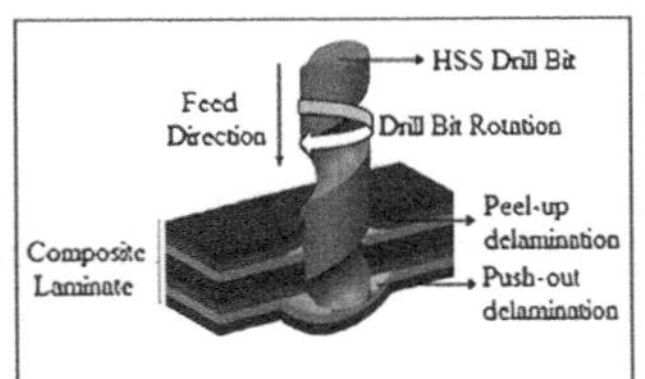

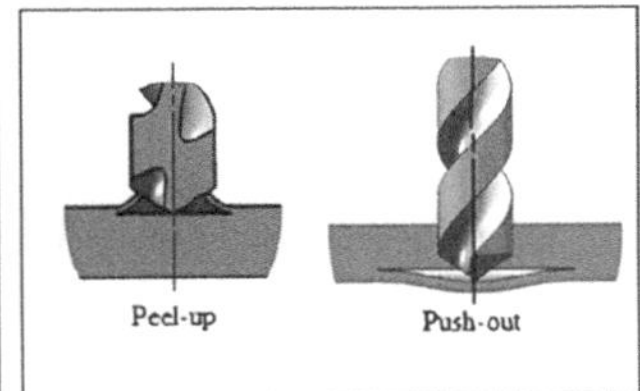

FIGURE 8.1 Delamination and other challenges in machining composites [24]

Fibre pullout: The fibres in composite materials can pull out during machining, which can result in poor surface finish and weaken the material.

To overcome these challenges, special cutting tools and techniques are often required for machining composite materials, such as using high-speed machining, coolant, and special tool coatings. Machining parameters, such as cutting speed, feed rate, and depth of cut, must also be carefully controlled to minimize damage to the material.

8.3 ENVIRONMENTALLY BENIGN MACHINING OF COMPOSITES

8.3.1 DRY MACHINING

Dry machining refers to the absence of cutting fluids during machining processes. To maintain efficiency, high-performance cutting tools are used in dry machining [25–27]. The advantages of dry machining are illustrated in Figure 8.2. It has become increasingly popular in manufacturing due to its environmentally friendly nature, worker satisfaction, compliance with laws and regulations, reduced coolant costs, and minimized leakage flow.

The machining of MMCs is very diverse from the machining of any normal material or alloy. The matrix and reinforcement phases, which the cutting tool continually contacts throughout the machining of MMCs, have an impact on the materials' machinability. MMCs may be machined using the standard machining technique under dry conditions. But while choosing the cutting instrument, caution must be exercised. For the purpose of machining MMCs, poly-crystalline diamond (PCD) cutting tools are thought to be more efficient [28]. Due to the reinforcement present, tool wear rises during the machining of MMCs, raising the overall cost of the operation [29]. An increase in tool wear, a poor surface quality, and resistance to plastic deformation are all effects of coarse reinforcement. Occasionally, during machining, reinforcing also contributes to the development of cracks on the matrix phase's shear plane [30]. The kind, size, and built-up edge formation of the reinforcing particles all affect the cutting force during machining [31]. The development of a built-up edge alters the actual rake angle, which alters the shear angle and directly influences the material's cutting force. The reinforcement pulls out as a result of weak bonding at the reinforcement-matrix contact, which increases the materials' surface roughness.

Additionally, researchers found that in-situ processed composites have finer reinforcement sizes and better bonding strengths than ex-situ MMCs [32]. Because of this, in-situ composites are machined with superior surface finishes. The cutting speed, reinforcement size, and volume fraction all affect how the chip is shaped when MMCs are being machined [33]. Extrinsic parameters like cutting speed, feed rate, depth of cut, and tooling system, as well as intrinsic ones like form, size, orientation, and volume percentage of the reinforcement, affect how machinable a material is [34]. Cutting force, surface integrity, chip shape, and tool wear mechanisms are all impacted differently by each parameter. By enhancing these machining parameters and tool geometry, the machinability of the MMCs may be increased [35]. So the dry machining of composites is not preferable.

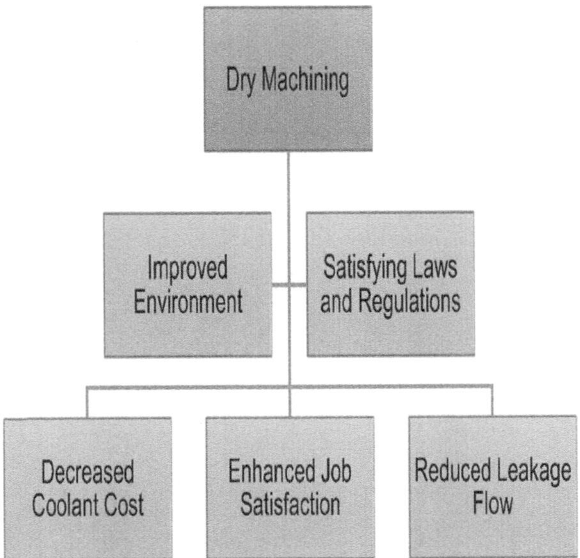

FIGURE 8.2 Benefits of dry machining [27]

8.3.2 CRYOGENIC MACHINING

Over the past 15 years, a lot of study has been done on the benefits of cryogenic machining in terms of its output quality and sustainability. Nevertheless, despite several authors experimenting with altered supply configurations for internal and exterior cooling of cutting tools, there is a research gap when it comes to optimizing cryogen supply factors for cryogenic machining. Additionally, it is discovered that cryogenic pre-cooling is a more efficient cooling technique than peripheral cooling of the machining area [36]. Figure 8.3 shows the applications of cryogenic machining.

8.3.2.1 Cryogenic Machining of Metal Matrix Composites

Attributable to the existence of hard ceramic particles in composites, particulate composites have difficulties with corrosion in machined components and rapid tool wear [36]. Despite these challenges, particulate composites particularly Mg based MMCs shows promise as structural materials for the automotive and aerospace sector because of its lightweight nature. Furthermore, the biodegradable feature of Mg-based particulate composites has made them a viable option for use in internal fixation implants in the biomedical field. To enhance the machinability of these particulate composites, some researchers have explored the application of cryogenic machining, which is elaborated upon in this section. All of these problems may be solved by using cryogenic fluids throughout the machining process as coolants and lubricants [37]. This section focuses on environmentally friendly cryogenic and hybrid machining methods that have been created to increase a material's machinability. Additionally, the usage of cryogenic fluid during machining is covered, along with the relevant input process parameters. In a thorough research of the impact of

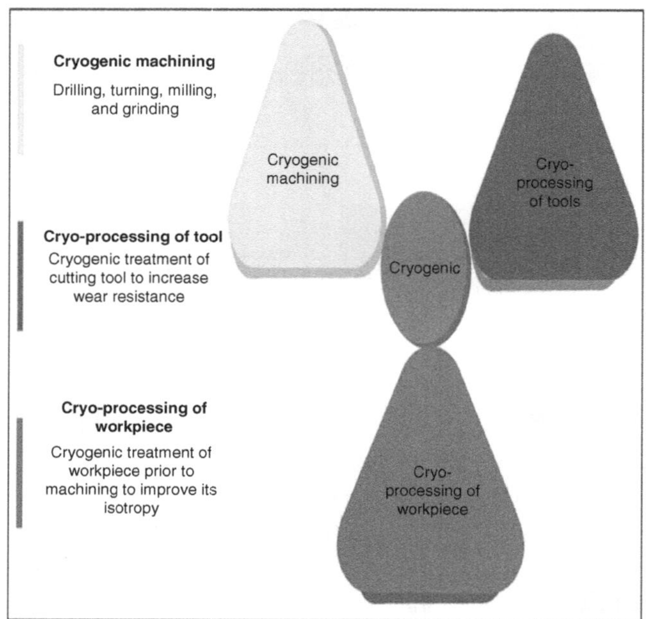

FIGURE 8.3   Cryogenic machining and processing application [35]

cryogenic machining on Mg-based particulate composites, Pu et al. [38, 39] looked at both surface integrity and corrosion tendencies in cryogenic environments.

A technique with setup comprising compressor, a cryogenic tank, a pneumatic pressure gauge, etc., is developed for supplying LN_2 to the cutting region during milling metal matrix composite (Al-5%TiCP). An LN_2 pressure controller, an L-shaped nozzle that was specially constructed, and a rotameter were all employed in the system (Figure 8.4). In comparison to dry and wet air conditions, the scientists found that LN_2 reduced temperature at the cutting zone, tool wear and improved the surface finish. Additionally, Khanna et al. [40] examined the differences in turning performance between the CUAT approach and the dry technique in Mg-based composite incorporated with SiC. According to Khanna et al. [41], the hybrid process was found to generate a better surface quality and chip breakability than the standard technique i.e. LN_2+MQL produced an even superior surface quality and chip breakability index than MQL and LN_2 procedures. While LN_2 and LN_2+MQL procedure used more power and cutting force, MQL approaches used less. The machinability of AXZ911/10SiC PMMC was examined in a different investigation using dry, LN_2, and LCO_2 machining processes [42]. It was found that dry condition produced a better surface finish than cryogenic conditions. LN_2 condition resulted in the least cutting force and power consumption when weighed against to dry and LCO_2 conditions.

8.3.2.2   Cryogenic Machining of Polymer Matrix Composites

The process of cutting fibre reinforced polymer (FRP) composites comes with a host of challenges, including tearing of fibres, delamination and fibre pullout in the

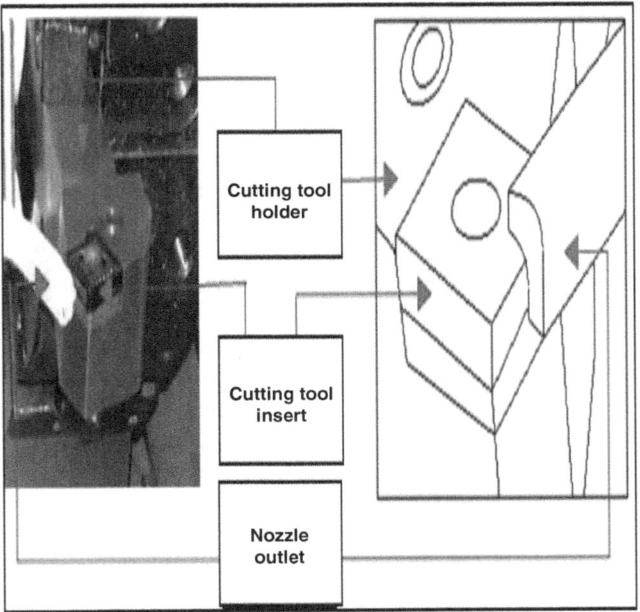

FIGURE 8.4 LN$_2$ delivery system [38]

resultant parts. To address these issues, some researchers have experimented with using cryogenic cooling of the polymer composites to reduce the differences in mechanical properties between the matrix and fibres. In a research on the machining of knitted FRP in a cryogenic atmosphere, it was discovered that this method significantly increased the machinability of the material [43]. The experiment contrasted applying liquid nitrogen (LN$_2$) continuously to the work piece vs heating LN$_2$ in a Dewar using an immersion heater. This method was found as more effective and over and over again produced a superior quality surface and improved life time of tool. It also generated enough pressure to pump LN$_2$ at 0.4–0.5 L/min flow rate. Similar results were observed i.e., better outcome for CFRP composites when drilling was done in a cold air environment (−10 deg), resulting in a longer tool life under changing operating circumstances [44]. Cryogenic shear cutting of FRP is recommended to resolve problems with delamination, distortion, and ripping of the fibres. The behaviour of the fibre reinforced plastics changed to more closely resemble that of a homogenous material as a result of the application of cryogenic fluid at low temperatures, which reduced the ductile movement of molecules in the matrix and may have made the resin matrix more brittle. Less harm was done to the cutting zone as a consequence [45].

Xia et al. [46] investigated the possibility of improving machinability of CFRP using a modified drill bit that delivers LN$_2$ to the cutting zone through two coolant supply holes. Despite the fact that cryogenic machining of CFRP increased delamination, it did lessen cutting edge rounding and outer corner wear, which led to better hole quality than dry machining. The machinability analysis on GLARE laminates discovered that the thrust force needed increased with both MQL and cryogenic environments. However, under MQL and cryogenic circumstances, the output surface quality was superior as opposed to dry conditions [47]. Demonstrated by SEM pictures, cryogenic circumstance increased the surface quality of the hole in rotary ultrasonic-assisted drilling procedures [48]. Basmaci et al. [49] advised utilizing small diameter drills while drilling CFRP composites, and their research revealed that drilling in a cryogenic environment boosted thrust force in comparison to drilling in a wet environment. The surface smoothness of the hole was enhanced by utilizing low-temperature drilling, nevertheless. Surface quality of the part increased while acoustic emissions and delamination reduced while drilling CFRP under cooled air settings [50]. Drilling of CFRP/Ti stacks under cryogenic environment is found more productive than wet drilling and that productivity may be increased by employing higher process parameters. Their findings, however, did not agree with those of previous investigations when it came to the reduced thrust force while utilizing cryogenic drilling [51].

A force-based model was proposed by Joshi et al. [52] to calculate delamination in dry and cryogenic drilling. In comparison to ordinary drilling, cryogenic drilling enhanced surface quality and tool life. The cutting force in cryogenic drilling has risen, as the tensile strength of CFRP composites has improved by 3% wherein less delamination and a better surface quality is observed [53]. When compared to dry drilling, cryogenic drilling produced much lower temperatures, smoother surfaces, and greater surface integrity. The enhanced reactions are being the result of a modification in the fibre fracture mode brought on by cryogenic drilling [54]. Utilization of cryogenic in drilling of a Ti/CFRP/Ti laminate composite resulted in decreased torque, damage to at interface, burr development while improvement in surface quality is achieved [55]. Under cryogenic drilling circumstances, a larger thrust force and reduced drilling-induced damage is observed in drilling of CFRP under cryogenic environment. Drilling in a cryogenic environment produces better-quality holes than dry drilling because it encourages brittle mode of fracture. Additionally, the drill bit with an angled tip reduced delamination better than the drill bit with a flat end [56]. The drilling capabilities of CFRP composites in dry and LN_2 conditions were examined by Khanna et al. [57]. When compared to dry drilling, they found that employing LN_2 drilling increased thrust force, enhanced hole quality, and decreased entrance delamination.

8.3.3 Machining of Composites with Bio-Based Oils

In recent years, the demand for vegetable oil-based metalworking fluids (MWFs) has increased significantly. This is because industrial rules are in place to protect the environment, prevent the use of dangerous mineral oil-based MWFs, and secure worker and worker well-being. Environmental and occupational health authorities are thus pressing the manufacturing sector to use MWFs in more sustainable ways. The superior functional qualities of vegetable oil-based MWFs have been highlighted in a number of studies [58–62]. These studies have also identified these MWFs as a viable substitute for hazardous mineral oil-based MWFs for cooling and lubrication needs. Vegetable-based oils are also utilized in EDM as dielectric fluids [63].

Bio oils can be used as a sustainable and eco-friendly alternative to traditional petroleum-based cutting fluids in the machining of composites. Bio-oils, also known as vegetable-based cutting fluids, are derived from natural sources such as vegetable oils, animal fats, and other renewable resources. They have been found to have several advantages over traditional cutting fluids, such as reduced environmental impact, improved biodegradability, and lower toxicity. In the machining of composites, bio oils can provide several benefits, such as improved tool life, reduced tool wear, and improved surface finish. They can also reduce the thermal load on the cutting tool and workpiece, leading to reduced machining temperatures and improved dimensional accuracy of the machined part. Furthermore, bio-oils can be used in combination with other additives such as lubricants, surfactants, and corrosion inhibitors to enhance their performance and provide additional benefits.

Overall, the use of bio oils in the machining of composites offers a sustainable and environmentally friendly solution that can improve machining performance while reducing the negative impact on the environment. There are several types of bio oils that can be used in machining composites, including: Soybean oil, sunflower oil, coconut oil, castor oil, palm oil, etc. Soybean oil is a popular bio oil used in machining composites due to its high lubricity and low viscosity. It is also readily available and cost-effective. Sunflower oil is another bio oil that is commonly used in machining composites. It has good lubricity and is resistant to oxidation. Coconut oil is a bio oil that is gaining popularity in machining composites due to its high lubricity, low viscosity, and good thermal stability. Castor oil is a bio oil that is commonly used in high-performance machining applications due to its high lubricity and ability to withstand high temperatures. Palm oil is a bio oil that is used in machining composites due to its good lubricity and low cost. It's worth noting that while bio oils offer several advantages over traditional petroleum-based oils, they also have some limitations. For example, they may not be suitable for all machining applications and may require more frequent changes due to their lower thermal stability. As with any lubricant, it's important to choose the right oil for the specific machining application.

8.4 SUMMARY

The chapter presented the fundamentals of composites, machining of composites, challenges in machining of composites and various possibilities of environmentally benign machining of composites such as dry machining, cryogenic machining, and usage of bio oils as lubricants. Due to the hard and abrasive nature of the composites, dry machining is not preferable though it is environmental friendly as it causes more heat generation, tool wear and inferior surface characteristics. On the other hand cryogenic machining with different additional setups were successfully utilized for effective machining of composites whereas hybridizing cryogenic machining with other techniques also yielded better results with minimal environmental impact. Usage of bio-based oils as cutting fluid in traditional machining and dielectric fluid in EDM is also fruitful in view of better output quality and minimal environmental effect.

REFERENCES

[1] Barnett HJ, Morse C. *Scarcity and growth: The economics of natural resource availability*. Routledge; 2013.

[2] Solow RM. *Sustainability: An economist's perspective*. Sel Readings Environ Econ; 1993.

[3] Short T, Lee-Mortimer A, Luttropp C, Johansson G. Manufacturing, sustainability, ecodesign and risk: Lessons learned from a study of Swedish and English companies. *J Clean Prod* 2012;37:342–352.

[4] Satheeshkumar V, Narayanan RG, Gunasekera JS. Sustainable manufacturing: Material forming and joining. In *Sustainable manufacturing processes*, edited by Narayanan, Ramabadran Ganesh and Jay Gunasekera, Elsevier; 2023, pp. 53–112.

[5] Velenturf APM, Purnell P. Principles for a sustainable circular economy. *Sustain Prod Consum* 2021;27:1437–1457.

[6] Allen D, Bauer D, Bras B, Gutowski T, Murphy C, Piwonka T, et al. Environmentally benign manufacturing: Trends in Europe, Japan, and the USA. *Int Des Eng Tech Conf Comput Inf Eng Conf*, vol. 80241, American Society of Mechanical Engineers; 2001, pp. 425–441.

[7] Panagiotopoulou VC, Stavropoulos P, Chryssolouris G. A critical review on the environmental impact of manufacturing: A holistic perspective. *Int J Adv Manuf Technol* 2021; 118:1–23.

[8] Sanchez LEA, Palma GL, Nalon LJ, Santos AE, Modolo DL. Different methods of cutting fluid application on turning of a difficult-to-machine steel. *Adv Mater Res*, Trans Tech Publ, 2013;628:476–481.

[9] Sanchez LE de A, Palma GL, Marinescu I, Modolo DL, Nalon LJ, Santos AE. Effect of different methods of cutting fluid application on turning of a difficult-to-machine steel (SAE EV-8). *Proc Inst Mech Eng Part B J Eng Manuf* 2013;227:220–234.

[10] Kazeem RA, Fadare DA, Ikumapayi OM, Adediran AA, Aliyu SJ, Akinlabi SA, et al. Advances in the application of vegetable-oil-based cutting fluids to sustainable machining operations—a review. *Lubricants* 2022;10:69.

[11] Dhar NR, Kamruzzaman M. Cutting temperature, tool wear, surface roughness and dimensional deviation in turning AISI-4037 steel under cryogenic condition. *Int J Mach Tools Manuf* 2007;47:754–759.

[12] Goindi GS, Sarkar P. Dry machining: A step towards sustainable machining–challenges and future directions. *J Clean Prod* 2017;165:1557–1571.

[13] Rajurkar KP, Sundaram MM, Malshe AP. Review of electrochemical and electrodischarge machining. *Procedia CIRP* 2013;6:13–26.

[14] Afiq Rashid M, Rahman M, Senthil Kumar A. A study on compound micromachining using laser and Electric Discharge Machining (EDM). *Adv Mater Process Technol* 2016;2:258–265.

[15] Jose M, Sivapirakasam SP, Surianarayanan M. Analysis of aerosol emission and hazard evaluation of electrical discharge machining (EDM) process. *Ind Health* 2010;48:478–486.

[16] Hatchett DW, Josowicz M. Composites of intrinsically conducting polymers as sensing nanomaterials. *Chem Rev* 2008;108:746–769.

[17] Kuzina E, Cherkas A, Rimshin V. Technical aspects of using composite materials for strengthening constructions. *IOP Conf. Ser. Mater. Sci. Eng.*, vol. 365, IOP Publishing; 2018, p. 32053.

[18] Kangishwar S, Radhika N, Sheik AA, Chavali A, Hariharan S. A comprehensive review on polymer matrix composites: Material selection, fabrication, and application. *Polym Bull* 2023;80:47–87.

[19] Balasubramanian M. *Composite materials and processing*. CRC Press; 2013.

[20] Gopal PM, Prakash KS. Minimization of cutting force, temperature and surface roughness through GRA, TOPSIS and Taguchi techniques in end milling of Mg hybrid MMC. *Measurement* 2018;116:178–192.

[21] Davim JP. *Machining: Fundamentals and recent advances.* Springer-Verlag; 2008.

[22] Wang SS. Fracture mechanics for delamination problems in composite materials. *J Compos Mater* 1983;17:210–223.

[23] Prakash KS, Gopal PM, Karthik S. Multi-objective optimization using Taguchi based grey relational analysis in turning of Rock dust reinforced Aluminum MMC. *Measurement* 2020:107664.

[24] Karataş MA, Gökkaya H. A review on machinability of carbon fiber reinforced polymer (CFRP) and glass fiber reinforced polymer (GFRP) composite materials. *Def Technol* 2018;14:318–326.

[25] Singh NK, Singh Y, Sharma A, Singla A, Negi P. An environmental-friendly electrical discharge machining using different sustainable techniques: A review. *Adv Mater Process Technol* 2021;7:537–566.

[26] Weinert K, Inasaki I, Sutherland JW, Wakabayashi T. Dry machining and minimum quantity lubrication. *CIRP Ann* 2004;53:511–537.

[27] Chandel RS, Kumar R, Kapoor J. Sustainability aspects of machining operations: A summary of concepts. *Mater Today Proc* 2022;50:716–727.

[28] Teti R. Machining of composite materials. *CIRP Ann* 2002;51:611–634.

[29] Byrne G, Dornfeld D, Denkena B. Advancing cutting technology. *CIRP Ann* 2003;52:483–507.

[30] Das B, Roy S, Rai RN, Saha SC. Study on machinability of in situ Al–4.5% Cu–TiC metal matrix composite-surface finish, cutting force prediction using ANN. *CIRP J Manuf Sci Technol* 2016;12:67–78.

[31] Tjong SC. Structural and mechanical properties of polymer nanocomposites. *Mater Sci Eng R Reports* 2006;53:73–197.

[32] Hung NP, Yeo SH, Lee KK, Ng KJ. Chip formation in machining particle-reinforced metal matrix composites. *Mater Manuf Process* 1998;13:85–100.

[33] Liu J, Li J, Xu C. Interaction of the cutting tools and the ceramic-reinforced metal matrix composites during micro-machining: A review. *CIRP J Manuf Sci Technol* 2014;7:55–70.

[34] El-Hossainy TM, El-Zoghby AA, Badr MA, Maalawi KY, Nasr MF. Cutting parameter optimization when machining different materials. *Mater Manuf Process* 2010; 25:1101–1114.

[35] Khanna N, Agrawal C, Pimenov DY, Singla AK, Machado AR, da Silva LRR, et al. Review on design and development of cryogenic machining setups for heat resistant alloys and composites. *J Manuf Process* 2021;68:398–422.

[36] Gopal PM, Prakash KS, Jayaraj S. WEDM of Mg/CRT/BN composites: Effect of materials and machining parameters. *Mater Manuf Process* 2018;33:77–84.

[37] Pu Z, Outeiro JC, Batista AC, Dillon OW, Puleo DA, Jawahir IS. Surface integrity in dry and cryogenic machining of AZ31B Mg alloy with varying cutting edge radius tools. *Procedia Eng* 2011;19:282–287.

[38] Josyula SK, Narala SKR, Charan EG, Kishawy HA. Sustainable machining of metal matrix composites using liquid nitrogen. *Procedia CIRP* 2016;40:568–573.

[39] Pu Z, Outeiro JC, Batista AC, Dillon Jr OW, Puleo DA, Jawahir IS. Enhanced surface integrity of AZ31B Mg alloy by cryogenic machining towards improved functional performance of machined components. *Int J Mach Tools Manuf* 2012;56:17–27.

[40] Khanna N, Suri NM, Agrawal C, Shah P, Krolczyk GM. Effect of hybrid machining techniques on machining performance of in-house developed Mg-PMMC. *Trans Indian Inst Met* 2019;72:1799–1807.

[41] Khanna N, Shah P, Suri NM, Agrawal C, Khatkar SK, Pusavec F, et al. Application of environmentally-friendly cooling/lubrication strategies for turning magnesium/SiC MMCs. *Silicon* 2021;13:2445–2459.

[42] Khanna N, Suri NM, Shah P, Hegab H, Mia M. Cryogenic turning of in-house cast magnesium based MMCs: A comprehensive investigation. *J Mater Res Technol* 2020;9:7628–7643.

[43] Bhattacharyya D, Allen MN, Mander SJ. Cryogenic machining of Kevlar composites. *Mater Manuf Process* 1993;8:631–651.

[44] Khairusshima MKN, Hassan CHC, Jaharah AG, Amin AKM, Idriss ANM. Effect of chilled air on tool wear and workpiece quality during milling of carbon fibre-reinforced plastic. *Wear* 2013;302:1113–1123.

[45] Watzke J, Schomäcker M, Brosius A. Cryogenic shear cutting of fiber reinforced plastics (FRP). *Procedia CIRP* 2014;18:80–83.

[46] Xia T, Kaynak Y, Arvin C, Jawahir IS. Cryogenic cooling-induced process performance and surface integrity in drilling CFRP composite material. *Int J Adv Manuf Technol* 2016;82:605–616.

[47] Giasin K, Ayvar-Soberanis S, Hodzic A. Evaluation of cryogenic cooling and minimum quantity lubrication effects on machining GLARE laminates using design of experiments. *J Clean Prod* 2016;135:533–548.

[48] Kumaran ST, Ko TJ, Li C, Yu Z, Uthayakumar M. Rotary ultrasonic machining of woven CFRP composite in a cryogenic environment. *J Alloys Compd* 2017;698:984–993.

[49] Basmaci G, Yoruk AS, Koklu U, Morkavuk S. Impact of cryogenic condition and drill diameter on drilling performance of CFRP. *Appl Sci* 2017;7:667.

[50] Abish J, Samal P, Narenther MS, Kannan C, Balan ASS. Assessment of drilling-induced damage in CFRP under chilled air environment. *Mater Manuf Process* 2018;33:1361–1368.

[51] Impero F, Dix M, Squillace A, Prisco U, Palumbo B, Tagliaferri F. A comparison between wet and cryogenic drilling of CFRP/Ti stacks. *Mater Manuf Process* 2018;33:1354–1360.

[52] Joshi S, Rawat K, Balan ASS. A novel approach to predict the delamination factor for dry and cryogenic drilling of CFRP. *J Mater Process Technol* 2018;262:521–531.

[53] Morkavuk S, Köklü U, Bağcı M, Gemi L. Cryogenic machining of carbon fiber reinforced plastic (CFRP) composites and the effects of cryogenic treatment on tensile properties: A comparative study. *Compos Part B Eng* 2018;147:1–11.

[54] Kumar D, Gururaja S. Machining damage and surface integrity evaluation during milling of UD-CFRP laminates: Dry vs. cryogenic. *Compos Struct* 2020;247:112504.

[55] Kumar D, Gururaja S, Jawahir IS. Machinability and surface integrity of adhesively bonded Ti/CFRP/Ti hybrid composite laminates under dry and cryogenic conditions. *J Manuf Process* 2020;58:1075–1087.

[56] Kannan S, Pervaiz S. Surface morphology of inclined CFRP holes when machined under cryogenic environment. *Mater Manuf Process* 2020;35:1228–1239.

[57] Khanna N, Pusavec F, Agrawal C, Krolczyk GM. Measurement and evaluation of hole attributes for drilling CFRP composites using an indigenously developed cryogenic machining facility. *Measurement* 2020;154:107504.

[58] Bennett EO. Water based cutting fluids and human health. *Tribol Int* 1983;16:133–136.

[59] Wickramasinghe KC, Sasahara H, Abd Rahim E, Perera GIP. Green Metalworking Fluids for sustainable machining applications: A review. *J Clean Prod* 2020;257:120552.

[60] Majak D, Olugu EU, Lawal SA. Analysis of the effect of sustainable lubricants in the turning of AISI 304 stainless steel. *Procedia Manuf* 2020;43:495–502.

[61] Krolczyk GM, Maruda RW, Krolczyk JB, Wojciechowski S, Mia M, Nieslony P, et al. Ecological trends in machining as a key factor in sustainable production–a review. *J Clean Prod* 2019;218:601–615.

[62] Benedicto E, Carou D, Rubio EM. Technical, economic and environmental review of the lubrication/cooling systems used in machining processes. *Procedia Eng* 2017;184:99–116.

[63] Yadav A, Singh Y, Singh S, Negi P. Sustainability of vegetable oil based bio-diesel as dielectric fluid during EDM process–A review. *Mater Today Proc* 2021;46:11155–11158.

9 Recent Developments in Environmentally Benign Machining

*Mustafa Günay, Ramazan Çakıroğlu
and Ahmet Tolunay Işık*

9.1 INTRODUCTION

Although machining methods are widely used in the automotive, aerospace, and defense industries, they carry many important environmental and occupational health risks such as energy or resource consumption, water pollution, air emissions, and health problems. New manufacturing technologies and management applications developed to minimize these problems are defined as green machining, sustainable machining and environmentally benign machining. On the other hand, it is important to address climate change concerns for manufacturing industries from large carbon emissions due to their high energy consumption and environmental impacts during the production phase of the lifecycle. In this context, an environmentally benign or sustainable production environment not only facilitates an organization by reducing the cost of machining but also contributes to improving machining efficiency and commercial competitiveness (Sen et al., 2021; Zhao and Sharma, 2015).

The continuous development and growth of the production industry bring along ecological and environmental damages such as the consumption of limited raw material resources, the spread of harmful emissions, and non-recyclable/non-degradable wastes. For this reason, it is important to consider environmental issues in production within the scope of standards such as ISO 14001 (Environmental management system) and ISO 14040 (Life cycle assessment) in terms of sustainability and efficiency of global competition. It is necessary to manage the use of raw materials and energy, temperature spread, toxic or harmful wastes released to the environment, processing quality, and total cost in accordance with the standards for minimizing environmental damage and for cleaner production. In this area, many studies have been carried out for environmentally friendly processing and therefore sustainability, and technological research continues to improve production processes and raw material used in terms of the health of the ecosystem. This chapter will enable a new generation of researchers to understand recent environmental trends in machining.

DOI: 10.1201/9781003352402-9

9.2 ENVIRONMENTAL EFFECTS OF MACHINING

Machining in the manufacturing industry is seen as a key process both in determining the quality of the end product and in terms of cost. Machining is a set of shaping processes in which the excess material is removed from the workpiece by various methods and thus the desired geometry and/or surface is obtained. It can be categorized as traditional, abrasive and unconventional machining methods. The conventional machining method is the method in which a sharp and harder cutting edge than the workpiece is used to cut the material mechanically. Conventional machining methods are divided into turning, milling, drilling and other machining methods (planning, broaching, etc.). Another method is abrasive machining, which removes the material mechanically with the effect of hard and abrasive particles. This method includes grinding and other abrasive manufacturing methods (Honing, lapping, etc.). In unconventional manufacturing methods, machining takes place using various forms of energy to remove material, without sharp cutting tools and abrasive particles. This manufacturing method consists of mechanical, electrochemical, thermal, and chemical forms of energy (Groover, 2010; Rohith et al., 2019). Various machining methods and types are classified in Figure 9.1.

Conventional machining is called the whole process of removing material from the workpiece with the use of sharp cutting tools. While the requirement for the hot hardness and wear resistance of the tool material to be higher than the workpiece, waste material (Chip) formation and the relatively long processing time create a disadvantage, the ability to process various geometric forms, high dimensional accuracy and low surface roughness offers the advantage of being used in the industry (Astakhov and Shvets, 2004; Stanojković and Radovanović, 2022; Williams et al., 2010). Mechanical energy is required during the machining by the cutting tool. Approximately 98%–99% of this energy used in chip removal is converted into heat, and 2%–1% is stored as elastic energy in the chip (Akhil et al., 2016; Luo et al., 2020). High temperature affects tool life, workpiece dimensional accuracy, and the operator.

Cutting fluids (Solid, liquid or gas) that serve as cooling and lubrication are generally used during machining in order to reduce the temperature and mechanical energy that will occur during machining (Figure 9.2). Coolants are fluids designed to remove heat generated in the cutting environment during machining. Lubricants are used in all sectors of modern industry to lubricate machinery and materials. Cutting fluids are used to decrease the machining temperature to increase productivity, to reduce tool/chip friction, and to extend tool life. Cutting fluids perform cooling, lubrication, cleaning, and rust prevention functions (Singh and Sharma, 2022).

The high energy used during processing, wastes in various forms and raw material consumption affect the ecological system negatively. Research indicates that aerosols released during processing cause various micro-bacterial diseases, respiratory tract infections, lung diseases, and skin damage. In addition, after long use of cutting fluids (Hydrocarbons, paraffin, organic compounds, oil-water emulsions, etc.), micro-bacterial activity, which is resistant to various disinfectants and heavy metals, increases on the counter. It is seen that disinfection does not completely destroy mycobacteria due to organic compounds and biofilm formation in cutting

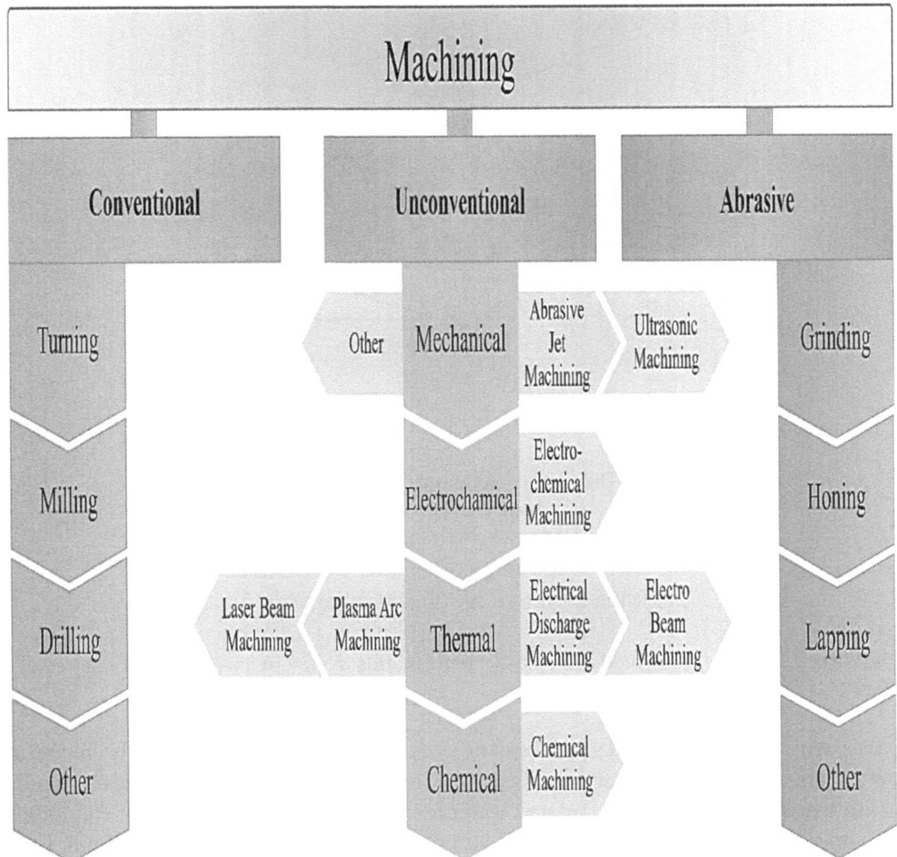

FIGURE 9.1 Machining methods

fluids. In order to reduce the micro-bacterial activity of post-processing waste cutting fluids and aerosols, cost-increasing solutions such as disinfection at high temperatures and filtration coated with hydrophobic compounds are suggested (Falkinham, 2003). Some studies have suggested the use of microbial cutting fluids, discovering that they can slow bacterial growth and increase processing efficiency at low cost (Teti et al., 2021) (Figure 9.3).

The need for advanced technologies in the processing of materials that cannot be processed with conventional methods has led to the development of advanced manufacturing technologies (Unconventional methods). The environmental effects of various methods used in the manufacture of hard or soft materials that are difficult to cut and process, from micro scale to macro scale, are examined by researchers (Bhattacharyya and Doloi, 2019; Bhowmik et al., 2019). The machining process can create hazards for the environment and the operator due to gas or toxic substances originating from the cutting tool material, auxiliary fluids used, or the workpiece. Methods such as electrochemical processing, chemical processing, electro-erosion

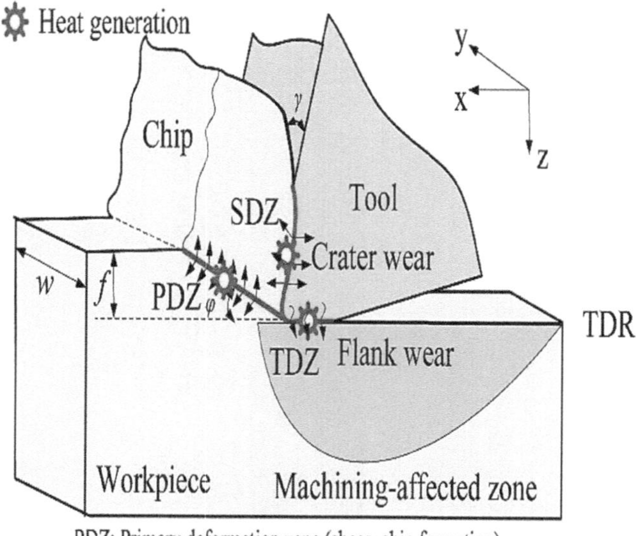

FIGURE 9.2 Heat generated and plastic deformation zones in machining (Liang et al., 2020)

processing and abrasive jet processing generally use environmentally harmful petroleum-derived chemical liquids in the manufacturing process. These liquids cause irreversible damage to soil and water resources as waste. Disposal of chemicals is seen as an economically sensitive and costly process. However, in methods such as laser processing, plasma arc processing and electron beam processing, gases and aerosols originating from the workpiece and cutting environment have dangerous effects on the health of the operator. These hazards cause chronic diseases in case of long-term exposure from short-term effects (Chandan et al., 2022; Gupta, 2020; Kiran et al., 2021). The electromagnetic field and ultrasonic frequencies released during ultrasonic processing have negative effects on human health such as hearing problems, muscle tension, increase in blood pressure, and hypertension (Radosz, 2020).

The equipment used during machining, from lubricating and cooling fluids to the workpiece, completely involves environmental effects. In minimizing these effects, it is important for the operator to apply occupational health and safety rules and to use environmentally friendly equipment during production. Less environmental impact; Economic production processes, which are efficient in terms of energy and resources, are accepted as the key to sustainable production. The processing of advanced materials with variable characteristics in many sectors poses a problem due to sustainability, energy consumption, and economic reasons.

Conventional cutting fluids are released to nature after use. The degradability of cutting fluids as waste is an important factor in preserving soil and water reserves. In

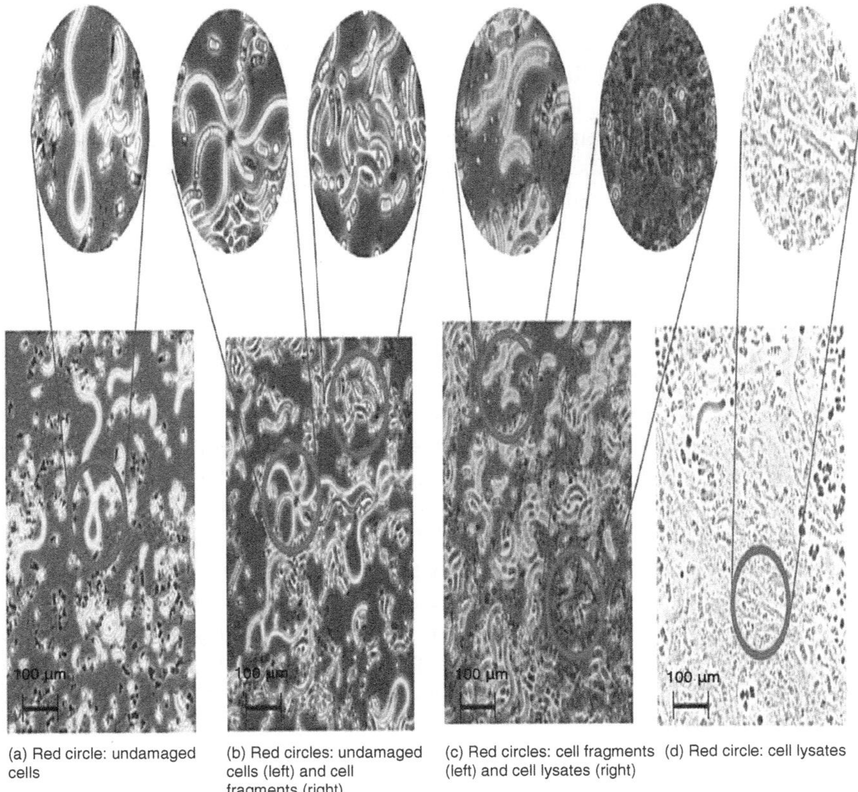

(a) Red circle: undamaged cells

(b) Red circles: undamaged cells (left) and cell fragments (right)

(c) Red circles: cell fragments (left) and cell lysates (right)

(d) Red circle: cell lysates

FIGURE 9.3 Microscopic view of the microbial cutting fluid (Byrne et al., 2021)

addition, aerosols and toxic gases formed with the high heat released during processing have a negative effect on the health of the operator. While asthma, cough, and other infections are predominant in those working in such environments, the risk of developing skin and lung cancer increases.

9.3 ENVIRONMENTALLY BENIGN COOLING AND LUBRICATION SYSTEMS

The manufacturing industry demands low cost, high quality and efficiency production. With appropriate machining techniques, high quality and productivity can be achieved at optimum processing parameters. However, high quality and efficiency should not include methods that will threaten environmental health (Bal and Erkan, 2019; Pimenov et al., 2022). Cutting fluids are used to prevent the heat, cutting forces and some negativity that occur during machining and to increase the quality of the process. Cutting fluids perform functions such as lubricating, cooling and removing chips in the machining area. Conventional cutting fluids are usually mineral oil based. Cutting fluids account for 10%–17% of the total production cost (Kui et al., 2022). Mineral oil-based cutting fluids are polluted by bacteria and fungi over time, and adversely affect the environment and human health if mixed with the soil (Kuram et al., 2013). For this reason, optimum use of cutting fluids according to ISO 14000 standards is important. When environmental cooling and lubrication systems are examined, the methods that are frequently used can be listed as flood cooling, dry machining, minimum amount of lubrication, pressure cooling, cryogenic cooling, and hybrid methods in which mineral oil alternatives are used (Dixit et al., 2012; Gupta, 2020).

In order to reduce and prevent the negative environmental impact of mineral or oil-based cutting fluids, the use of water-oil emulsions, synthetic, semi-synthetic, and vegetable or animal based oils is becoming widespread (Sankaranarayanan et al., 2021; Tang et al., 2022). Although synthetic oils have a long service life, they cause some problems in terms of cost. Vegetable and animal-based oils, which were used as cutting fluid before mineral oils, are now being used again due to environmental effects (Figure 9.4). These oils can be produced easily and cheaply with high lubricating ability. However, vegetable oils have some shortcomings such as low cooling ability, deterioration times and risk of fire. It is generally used as a cutting fluid at low cutting speeds in machining operations where high lubrication needs are required. Alternatively, water-based cutting fluids can be used at high cutting speeds as they do not pose a fire risk. However, the use of water-based cutting fluids is limited due to corrosion and rust effects (Benedicto et al., 2017; Kazeem et al., 2022; Kumar Gajrani and Ravi Sankar, 2017). As a result, other alternative options such as dry machining, MQL, various emulsions, nanoparticle-based cutting fluids should be preferred. These options are environmentally friendly and cost-effective approaches.

Minimum quantity lubrication (MQL) method can be used in order to reduce the cost of using cutting fluids and to prevent these problems. In the MQL method, the lubricating fluid is sent to the cutting zone with the help of compressed air. Tool friction is reduced by the lubricating fluid, while compressed air helps remove chips. However, with the MQL method, reaching the cutting zone of the lubricating fluid is limited at high cutting speeds. Therefore, the processing of hard materials that are difficult to machine at low or medium cutting speeds increases the demand for the MQL method (Baron et al., 2016; Sharma et al., 2016).

MQL application helps in optimization of parameters, lowest tool wear, roughness, cutting forces, and energy consumption. It provides significant increases in efficiency under optimum conditions. For example, edible vegetable oils such as peanut, walnut, soybean, sunflower, mustard, cottonseed, and coconut oil are used successfully as cutting fluids. However, inedible oils such as castor, jatropha, gingelly, karanja, and neem oil have proven to be excellent choices as cutting fluids considering food safety. How a vegetable oil can be beneficial depends on the parameters such as workpiece and tool material. In its pure form, emulsified vegetable oils and mixing different vegetable oils are useful options during processing. The thermal conductivity of vegetable oils is improved by adding nanoparticles with high thermal conductivity. Silver, boric acid, graphite, carbon, and molybdenum disulfide (MoS_2) nanoparticles were used as additives. High concentrations of nanoparticles affect the surface finish by infusing the nanoparticles onto the surface and being sheared

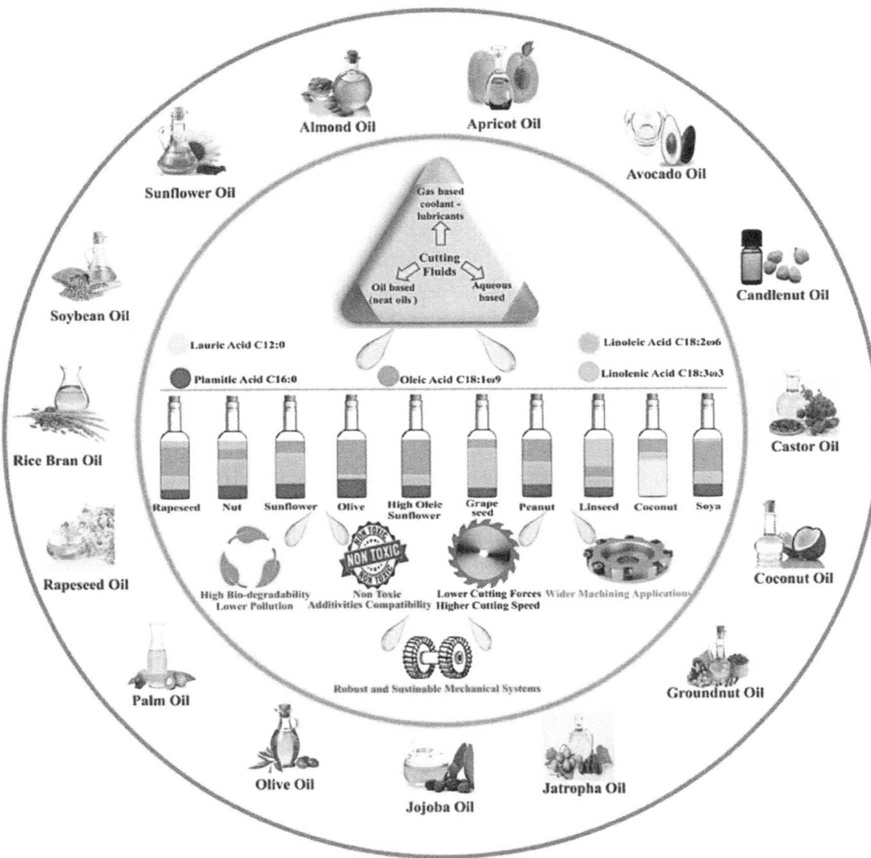

FIGURE 9.4   The effect of vegetable oil-based cutting fluids on machining (Sankaranarayanan et al., 2021)

by oncoming particles. Rolling the nanoparticles formulates a lubricating film that can be easily cut, thus providing well surface finish. The most convenient way to use vegetable based cutting fluid is to supply their nanofluids in sufficient concentrations to the machining zone with the MQL. Also, the focus should be on the choice of inedible vegetable oils.

Concern about the environmental health effects of cutting fluids makes dry machining more and more feasible. Dry machining is a method in which material removal machining is performed without cutting fluids (Goindi and Sarkar, 2017). Research reveals that it is difficult to catch the advantages of dry machining, such as improving the tool life and surface quality of cutting fluids, and removing the high heat generated during cutting. However, dry machining is seen as an economic and feasible method that reduces the cost of cutting fluids. The high heat that will occur during dry machining triggers friction and sticking, negatively affecting tool life and workpiece quality (Debnath et al., 2019). However, it was reported by the researchers that the effect of rapid cooling during dry cutting, thermal shock and the negative effects of tool wear during step cutting were eliminated (Jain and Kansal, 2017). To contribute to the sustainability of dry processing; Many methods have been developed such as 1- cooling through the tool, 2- continuous cooling of the workpiece from a region that does not reach the cutting zone, 3- continuous cooling of the cutting medium using cryogenic systems (Ramachandran et al., 2017; Singh et al., 2016).

High temperatures during cutting can reduce tool life and workpiece quality. Cryogenic processing, which is used as an alternative, is the cooling application performed at very low temperatures (Below zero) with the help of some gases such as nitrogen, oxygen, helium, hydrogen. Generally, the cryogenic method is used in materials that are difficult to process such as Ti alloys and superalloys and that generate high temperatures during processing (Yap, 2019). With the cryogenic method, it is aimed to reduce high cutting temperatures and to cool the tool-chip interface. Cryogenic fluid can be applied directly or indirectly to the cutting area, workpiece surface, through the cutting tool with different methods. In this context, it is used as a solution to the application cooling problem, increasing the metal removal rate, obtaining good surface quality, and providing efficient, environmentally friendly and low cost machining by providing longer tool life. In addition to its advantages, cryogenic machining can cause thermal shocks, crack propagation, chip adhesion, and tool damage due to intermittent contact of the tool teeth with the chip surface in milling. Difficulties such as the cost of installation, post-use recycling of cryogenic fluids, and providing appropriate combinations of processing cooling parameters limit the use of this method (Shokrani et al., 2016). Figure 9.5 shows the application of cryogenic cooling and MQL together.

It has been observed that cooling and lubrication methods have deficiencies in the application areas. In order to eliminate these deficiencies, hybrid methods have been developed. Research shows that the use of methods such as nanofluid-assisted machining, MQL-cryogenic, dry-cryogenic, solid lubricant–assisted machining is appropriate in terms of processing quality, cost, and environmental health (Hoghoughi et al., 2022; Khan et al., 2020; Korkmaz et al., 2021; Ross et al., 2021; Song et al., 2020). Figure 9.6 shows nanofluid-assisted processing as a hybrid method.

In this context, the impact of production processes and systems on the environment, economy, and society is the determinant of sustainability. It is important for the environment and sustainability to choose methods that protect energy and natural resources, and produce safe, economic, and robust products for employees and consumers. When the cooling and lubrication technologies for environmental manufacturing are examined, the comparison of their performance and environmental effects is given in Table 9.1.

9.4 RECENT RESEARCH ATTEMPTS

9.4.1 Conventional Machining

Cooling and lubrication constitute an important part of the total production cost in machining. Fluids such as mineral oils and gases that can provide cooling and lubrication should be used to reduce friction between the cutting tool and the workpiece,

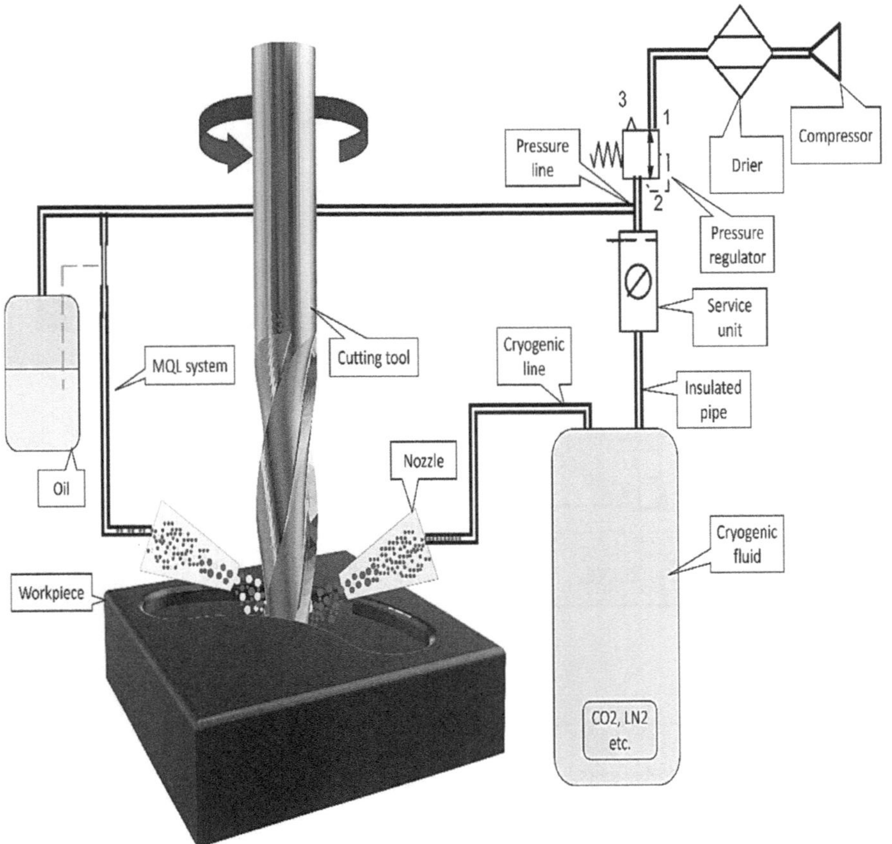

FIGURE 9.5   MQL and Cryogenic cooling application

reduce the cutting forces and temperature, and prevent BUE (Built-up-edge). The cost, emission, degradability, and waste of the fluid used are important criteria for sustainability. In this context, recent studies on the effects of cutting fluids used in conventional machining are presented below.

The increasing cost of petroleum and mineral-based cutting fluids due to environmental damage and limited raw material resources has also limited its use. The MQL method, which is used to minimize these disadvantages, is realized by sending pressurized gas and lubricating liquid to the tool/chip interface. By providing sufficient lubrication on the pressure-advancing aerosol processing surface, it minimizes the harmful effects of cutting fluid and prevents its excessive use (Çakıroğlu, 2022). Günay et al. analyzed the effects of dry, air-cooling conditions, and aerosol containing vegetable-based oil (5%) mixed with five bar compressed air (95%) on tool life and machined surface while turning Nimonic 80A super alloy using coated carbide tool. The authors mentioned that the oil spray method provides less tool wear compared to other methods. In addition, it was stated that the

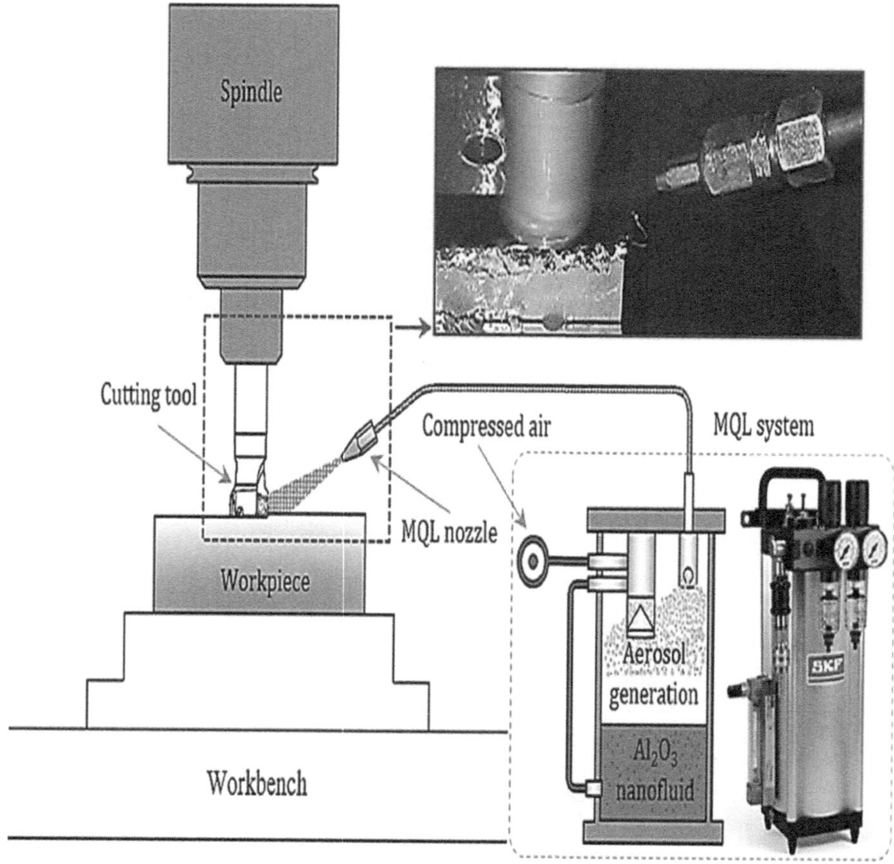

FIGURE 9.6 Nanoparticle fluid-assisted machining (Günan et al., 2020)

microhardness of the workpiece decreased as the machining depth increased, and increased in oil, air and dry cutting regimes, respectively (Günay et al., 2020). It is said that AISI D2 steel shows high performance (lower cutting force, better surface quality, and machining temperature) like the wet method, while MQL keeps the cutting area clean as in dry grinding in dry, wet (Synthetic oil), and MQL environments (Khan et al., 2018). Yin et al. performed milling experiments on AISI 1045 steel using cottonseed, palm, castor, soybean, and peanut oil-based MQL. The authors noted that palm oil provides the best frictional resistance and shear force. In addition, vegetable oils with high saturated fatty acids (25.12, 49.9, 1.36, 15, 21 respectively) are ideal for reducing shear force and friction coefficient by forming a stable and wide tribofilm layer. A certain amount of antioxidant supplementation is recommended for vegetable oils in order to slow down the oxidation of unsaturated carbon bonds.

Alternatively, biodegradable and temporarily environmentally harmful vegetable-based fluids (oils) can be used as cutting fluids at low cost. Zahoor et al. investigated

TABLE 9.1

Comparison of cooling/lubrication regimes

Method	Advantage	Disadvantage
Mineral oil	Lubrication	Poor thermal conductivity
	Rust and corrosion resistance	Low flash point
	High diffusion	Negative impact on the environment and human health
	Long lasting	Limited raw material resource
Vegetable oil	Biodegradability	Production cost
	Environmentally friendly	Oxidization
	High flash point	Poor corrosion resistance
	Lower costs	Low thermal stability
	Low toxicity	
Solid lubricant	High thermal stability	Friction and wear during lubrication
	Corrosion resistance	Storage shortage and lifetime
	Durability	Pollution
	High speed operability	
	Permanent lubrication	
MQL	Low lubrication cost	Effective lubrication
	Effective cooling ability	Evaporation of droplets during processing
	Low cost of use	Pollution from the use of petroleum oils
	Wear reduction	
Cryogenic	Ecofriendly	Low cutting speed
	Minimizes tool wear	High cost of use
	Lubricity and low burr formation	Equipment problem
Dry	Environmentally friendly and recyclable	Low cooling and no lubrication
	Low in pollution and toxicity	Failure mechanism
	High washing ability	Filter and dryer equipment maintenance
	Chemically inert	
Hybrid	High stability	Storage shortage
	Good lubrication and cooling	Cost of use
	Better thermal conductivity	Environmental impact
	Work balance	Disposal and treatment problems

the effects of cutting speed (65–95 m/min), feed per tooth (0.10–0.20 mm/tooth) and depth of cut on MRR, flank wear and surface structure while milling Inconel 718 using synthetic vegetable ester-based coolant. The authors noted that the maximum flank wear of the tool was 101.50 μm, MRR with increasing feed and depth of cut, and Ra increased with increasing feed and cutting speed (Zahoor et al., 2020). Gunjal and Patil evaluated the effects of vegetable cutting fluids (Canola, coconut and soybean oil) on surface roughness, tool life and wear, in comparison with dry and synthetic oil media, during the MQL method when turning AISI 4340 steel. The authors, who stated that synthetic oil showed better tool life performance at high cutting speeds, stated that canola oil with higher density (962 kg/m^3) and heat transfer coefficient (251.4 W/m^{2}K) provides better tool life in the vegetable cutting fluids category. While lateral wear is faster at increasing cutting speed (200–240 m/min), sudden changes in surface roughness are not observed with constant feed rate (Gunjal and Patil, 2018).

Dry or gaseous processing method can be applied in order to eliminate the negativities of cutting fluids. With some dry processing techniques (Under-cooling, internal cooling, cryogenic and thermoelectric cooling system), the removal of sawdust from the processing area and cooling can be achieved. However, cutting parameters need to be optimized and adapted to dry conditions. Dry processing in the industry is seen as a method that reduces production costs and protects the environment and operator health. Legislation and the implementation of new standards increase the demand for dry machining for economic and environmental reasons (Sreejith and Ngoi, 2000). The tool, which is deformed by the high temperature formed during dry machining, causes the formation of thin and strip-like chips, affecting the surface quality and dimensional accuracy. However, reducing the high temperature generated during dry processing can be achieved by applying cryogenic cooling (Çakıroğlu, 2021). However, in this case, sudden temperature changes have a negative effect on thermal shock resistance, as in cutting fluids used for cooling. In a study examining the wet and dry turning of LM 25 aluminum alloy comparatively, it was stated that the wet condition had a relatively better effect on the performance results (MRR, Ra and circularity) (Dhanalakshmi and Rameshbabu, 2021). Li et al. evaluated the microstructure changes and chip surface oxidation in dry milling of AISI H13 steel. The formation of various iron oxides such as FeO, Fe_3O_4 and Fe_2O_3 on the chip surface has been mentioned because about 96% of the thermal energy is transferred to the chip surface during machining and dry machining takes place at high temperatures. Maier et al. investigated the effect of tribo-oxidation on the tool structure as a result of dry milling of Ti6Al4V alloy in laboratory air and oxygen-free (XHV) conditions. They provided an oxygen-free environment with the 1.5% SiH4 doped argon gas system sent between the spindle and the flexible plastic casing fixed to the base plate. It was observed that the cutting forces decreased and the adhesion was less in the processes carried out in this environment. Researchers have mentioned the positive change in chip structure in the XHV environment (Maier et al., 2020).

In terms of machining performance, cutting fluids should have properties such as low viscosity, which provides better lubrication, and high thermal conductivity, which provides rapid heat transfer. However, vegetable oils recommended for sustainable manufacturing have poor thermal conductivity. Therefore, suitable concentrations of nanoparticles such as graphite (0.2663 W/mK), MoS_2 (0.2362 W/mK) and Al_2O_3 (0.2085 W/mK) with high thermal conductivity should be used for improving the thermal conductivity of vegetable-based cutting fluids (0.1432 W/mK) (Gupta et al., 2019). Many studies have shown improvement in machining (Tool wear, cutting force, surface roughness and power consumption, etc.) and tribological performance with the use of nanoparticles in cutting fluids (Cetin and Kabave Kilincarslan, 2020; Kumar Sharma et al., 2020; Yi et al., 2019). Hegab et al. used nanoparticle reinforced (0/2/4) vegetable-based cutting fluid in the machining of Ti6Al4V alloy with MQL. It has been reported that due to the formation of a tribofilm layer at the tool-chip interface, the tool wear is reduced and the surface roughness is improved by 38%–50% (Hegab et al., 2018a). Another study mentions the positive effect of nanoparticles on tool wear and power consumption (Hegab et al., 2018b). In a study using vegetable, mineral and vegetable based MoS_2 or calcium fluoride (CaF_2) nanoparticle added oils, turning experiments were carried out with the MQL method. Less adhesion, better friction coefficient, surface roughness and cutting force results were obtained in machining with MoS_2 added MQL, which has better thermal conductivity, specific heat and viscosity than other liquids (Gajrani et al., 2019). Singh et al. used carbon nanoparticle added soybean oil (NMQL) as the cutting regime in turning Inconel 625 alloy. Better surface quality (0.493 µm) and tool flank wear (146.67 µm) compared to dry machining of the NMQL method, and lower surface roughness (5.48%) compared to the under flood coolant method were obtained (Singh et al., 2018). Wang et al. investigated the grinding performance of Inconel 718, ductile iron and AISI 1045 materials with palm oil with nanoparticles (MoS_2 and Al_2O_3). The researchers stated that MoS_2 nanofluid should be used for better machining quality in medium carbon steels and lower relative wear in cast irons, Al_2O_3 nanofluid should be used for better machining quality in nickel-based alloys and better strength ratio in cast irons (Wang et al., 2018). On the other hand, the use of nanoparticles in cutting fluids may cause some environmental problems. In order to avoid this situation, the waste nanofluid must be carefully filtered and purified from toxic particles (Wang et al., 2020).

In past studies on environmental machining techniques, the problem of tool wear is one of the reasons that affect machining performance, power consumption and product quality. In order to realize sustainable production, different techniques have been tried as a solution to this situation and positive results have been obtained. Salguero et al. tested the tribological wear mechanism using simple cemented carbide pins for dry machining of Al-Cu alloy. The obtained results show the workpiece adhesion after tool wear with temperature and the effect of tool particles on wear (Salguero et al., 2018). Uddin et al. investigated the machining performance of coated (PCD, TiAlN, TiN) and uncoated carbide tools in dry turning of Al 2024 alloy. The authors reported that the TiAlN coated insert exhibited less tool wear (Min 14 µm) but higher adhesion than other tools. In addition, it has been stated that the debris formed by the wear of the TiN coating affects the lubrication mechanics and causes an improvement in the surface roughness of the workpiece (Uddin et al., 2021). It has been stated that, in the turning of Ti6Al4V alloy, double-textured (groove and pit) cutting tools reduce the cutting, feed forces, and friction coefficient by 16.2, 31.2, and 17.8%, respectively. The reduction in adhesion and edge formation was attributed to the wear residue trapped in the tissues (Siju et al., 2021). Likewise, it is mentioned that the micro-textured tool produced by using electro-discharge technique on HSS in turning Al 7075-T6 alloy reduces the cutting forces (13% at 20 m/min speed, 5% at 30 m/min speed compared to nonwoven tool). In the study, as a result of filling the cavities in the surface textures with MoS_2 solid lubricant, the tool edge adhesion decreased to 0.821 mm and a 10% reduction in cutting forces was obtained at 30 m/min (Singh et al., 2019). Cui et al. performed the intermittent turning process with the tool textures they created with bio-inspired design (Cui et al., 2020). Jamil et al. evaluated the machining time, cutting energy, energy efficiency, and carbon emission contents in the milling of Ti6Al4V alloy in dry, alcohol-based MQL, LN_2, CO_2 environments. While the lowest and highest machining time was performed in CO_2 and dry machining environment, the difference in carbon emission improvement compared to dry machining was in CO_2 (4.58%), LN_2 (3.57), MQL (1.28) environment, respectively. Also, the minimum cutting temperature (110°C) LN_2 In the dry machining environment, the reduction in side wear (54, 42, 24%) was in CO_2, LN_2 and MQL, respectively. From the results obtained, the researchers emphasized that the CO_2 environment exhibits the best sustainable performance (Jamil et al., 2021).

As seen in recent studies, each method has some disadvantages. In order to eliminate these, more research on environmentally friendly processing technologies is required, and it will also be more meaningful to work on the optimization of process parameters and processing simulations based on material models.

9.4.2 RECENT WORK IN UNCONVENTIONAL MACHINING

The need for different manufacturing methods industrially has led to the emergence of unconventional machining methods. Of these methods, electrical discharge machining, abrasive water jet machining, chemical machining, laser beam, and ultrasonic machining are frequently used. Each has its own unique properties for precision machining of engineering materials and making geometries and micro-parts that are difficult to manufacture using conventional methods. However, the use of certain fluids and consumables, toxic waste, pollution, thermal damage, toxic fumes, and gases are the main environmental problems associated with these processes. Many environmentally benign machining strategies are being developed to cope with these challenges and contribute to sustainability.

Despite the numerous advantages of electrical discharge machining (EDM), there are some limitations. For example, huge energy consumption and slow material removal rate, as well as dangerous emissions, toxic dielectric and slurry formation can lead to various health and environmental problems. To overcome these problems, environmentally friendly processing is recommended by applying dry or near-dry EDM with vegetable, water and gas-based bio-dielectric fluid (Kalyon and Fatatit, 2019; Singh and Sharma, 2017). In addition, hybrid EDM processes such as magnetic field-assisted EDM, ultrasonic and rotary ultrasonic-assisted EDM, ultrasonically assisted cryogenic cooled electrode-assisted EDM, abrasive, and rotary abrasive jet-assisted EDM, and ultrasonic-assisted waterjet machining EDM consume less energy and can improve output parameters or responses such as surface integrity, material removal rate (Arif et al., 2022).

The abrasive machining method is the process of removing chips from the workpiece surface as a result of sending abrasive particles at high speed with the help of air (AAJM) or water (AWJM) jet. However, the transport of abrasive particles in air and their scattering after processing poses a problem for environmental safety (Melentiev and Fang, 2018). Alternatively, the abrasive waterjet processing method offers a more environmentally friendly approach that does not produce toxic fumes during processing. With this method, chemicals that will endanger the health of the operator are not formed during processing and the water does not pollute the environment as waste. However, there are disadvantages such as high geometric deviations, burr and crack formation, rust and corrosion effects on metal parts, high noise and high machining cost in case the parameters are not selected appropriately in the abrasive water jet machining method. Most abrasive materials used in jet processing are harmful to the environment as waste and are costly to recycle. Perec, in its research on the breaking and fragmentation of abrasive grains using the AWJM method, has achieved an average of 45% recycling potential in recyclable abrasives made of natural mineral stones. He also mentioned the usability of shredded abrasives after recycling by creating sharp corners (Perec, 2021). Jet processing using ice as an abrasive particle has provided a solution to the cost of recycling and storing abrasive particles as waste. Jerman et al. observed cryogenically cooled ice particles in a high-velocity water jet (Max 200 MPa, 633 m.s^{-1}), and reported that it caused similar damage to the garnet abrasive on glass and aluminum materials. It has been emphasized that ice particles cause larger volume craters and high metal removal rates can be achieved with this method (Jerman et al., 2021).

Ultrasonic machining (USM) is basically the process of removing material from the workpiece by the ultrasonic vibration of abrasive particles from a rotating or stationary tool. It is sent to a slurry processing area with abrasive particles, and after the vibration of the tool, cavitation or chemically removes chips from the workpiece (Saptaji et al., 2018). This method, which is generally used in the processing of ceramics and biocompatible materials, investigates the effect of ultrasonic vibrations on human health and the environment of abrasive slurry. Singh and Sharma investigated the sustainability effect of ultrasonic-assisted grinding of Nimonic 80A material. They aimed to measure the grinding performance and environmental effects of cutting fluid. The results showed that green atomized coolant improves grinding performance and reduces environmental impacts (Singh and Sharma, 2022).

Chemical machining is the process of controlled corrosion of the workpiece by means of a tool in the chemical abrasive pool. Here, in order for the chip removal mechanism to be active, the workpiece material must be able to react with abrasive chemicals. This situation causes dangerous toxic wastes and harmful gases for health. It is necessary to carry out the manufacturing processing taking into account the environmental and health effects of the solution and workpiece to be used (Clarke et al., 2018). Zhang et al. investigated the properties of the chemical developed in the polishing of copper by chemical processing and its effect on the processing. By using corrosive chemicals containing environmentally friendly silica, H_2O_2 and COS, they obtained better surface quality (Ra = 0.444 nm) than sulfuric acid and citric acid environments (Zhang et al., 2019). Electrochemical machining (ECM), on the other hand, is a thermal and chemical metal removal process by transferring an electrical charge between two electrically conductive materials. In the process, which is carried out in an electrolytic liquid, salt water solution can usually be used due to environmental effects. The brine solution used in the ECM can cause rust and corrosion on the workpiece surface, endangering the health of the operator (Rajesh et al., 2022). Absence of tool wear compared to conventional machining methods and even EDM, and obtaining stress-free and crack-free surfaces even in parts with complex geometries are among the most important advantages. ECM are used to develop a pulse/pulse reverse style for electro polishing and etching as well as applications such as gears, fluid control valves, stents, and complex cavities. As previously mentioned, ECM applications include aerospace, biomedical, deburring, energy, automotive deep hole machining, and tribology.

9.5 ENERGY CONSUMPTION IN MACHINING

Conventional machining is a process in which production is made by converting electrical energy into mechanical energy. In unconventional machining, however, three main forms of energy use thermal, chemical, and electrical energy. As a result, the main form of energy used in all machining methods is electrical energy. In addition, considering that more than 80% of the energy used is obtained from fossil fuels such as coal, oil, and natural gas, it is known that harmful gases released into the atmosphere cause very bad effects for both human health and the environment. In this context, energy saving in the processing process is attracting more and more attention, according to the requirements of sustainability and environmental awareness in the manufacturing industry. Energy saving methods in conventional and unconventional machining often focus on processes such as machine tool design, modeling of machining parameters, monitoring of machining status, and designing and optimizing key components (Zheng et al., 2020). However, the integration of process scheduling and production arrangement for low-carbon manufacturing helps both to optimize energy consumption in the manufacturing process, and to improve the shop floor's manufacturing operations (Kong et al., 2022). Today, energy demand is constantly increasing with the increase in population and the development of technology. For this reason, it is very important to produce the energy used from renewable energy sources and to reduce the amount of energy used. While the industrial sectors use approximately 37% of the energy produced in the world, 90% of the energy consumed through the production processes in the industrial sector, and 36% of the CO_2 emissions belong to the machine tools (Yusuf et al., 2021). For this reason, the manufacturing industry has had to not only meet customer needs, but also reduce energy consumption and develop environmentally friendly and sustainable production methods (Daniyan et al., 2021).

While time, quality, energy used and cost are important in sustainable manufacturing, the improvement of harmless processing conditions for persons and nature comes to the fore in environmentally benign machining. Machining parameters in conventional machining directly affect cutting forces, cutting zone temperature, surface roughness, tool wear and energy consumption. Increasing demand for energy day by day increases the importance of selecting suitable processing parameters and saving energy (Zhong et al., 2017). In addition, innovative and environmentally benign cooling methods are developed to improve the quality, efficiency, productivity, and sustainability of the materials (Agrawal et al., 2020; Dornfeld, 2014). Reducing cutting fluids, which are used in large quantities in machining by new methods, make a countless contribution to both the health of employees and the protection of natural resources. For this reason, recent studies on the production of parts with desired properties with processing conditions that are both environmentally friendly and minimize energy consumption are summarized below.

In the machining process, reducing product cycle time helps reduce energy consumption and associated emissions. Hydraulic equipment used to send coolant and oil from the machine tool to the cutting zone consumes a significant amount of electrical energy. Even optimizing the pump quantity and the coolant quantity results in a reduction in energy consumption. In this regard, replacing conventional flood cooling with MQL or MQCL techniques results in less electrical energy consumption (Pervaiz et al., 2020). At the same time, a lot of research is carried out to determine the effects of processing parameters on energy consumption, to reduce and estimate the amount of energy consumption (Jia et al., 2017; Lv et al., 2018) XX. In a study on carbide drilling of VT-20 Ti alloy, the cryogenic medium using LN_2 was found to be the most economic and ecological due to the larger number of holes drilled with relatively lower impact on human health, ecosystem and natural resources compared to flooding and LCO_2 cutting environments (Khanna et al., 2021). Agrawal et al. investigated energy consumption, surface quality and machining costs in turning Ti6Al4V with newly developed environmentally friendly hybrid techniques. In terms of machine tool energy consumption, cryogenic-ultrasonic-MQL-assisted and cryogenic-ultrasonic-assisted turning showed an improvement of 1.8%–3.6% compared to cryogenic machining (Agrawal et al., 2020). It turns out from the literature that energy and metalworking fluid consumption reduction can be achieved by using the most suitable cooling/lubrication techniques, machining parameters optimization and advanced machining tools.

9.6 CONCLUSIONS AND FUTURE WORK

While time, cost, quality and energy consumption are important in sustainable processing, it is important to minimize the harms to humans and nature in environmentally benign processing. MQL, dry, cryogenic or hybrid cutting regimes should be used in order to reduce energy consumption as well as to prevent operator and environmental damage caused by the use of ordinary cutting fluids in conventional machining. The use of vegetable oils instead of mineral-based oils in MQL applications is important for the protection of soil and water resources as waste. Although it is predicted that the cryogenic method will provide a safe processing, it is seen that it is lacking in machining performance and reducing the friction forces that occur during machining. Although dry machining is seen as the most environmentally benign method, this method brings many problems such as end product quality, increase in cutting temperatures and reduction in tool life. Research on tool material and geometry to improve machining performance, along with the development of coating technologies, has led to longer tool life, especially when machining difficult-to-cut materials. However, although these developments in tool producing processes increase machining costs, it has been emphasized that productivity will increase when accurate process planning and optimization of machining conditions are provided.

In applications where conventional methods are insufficient, unconventional machining methods come into play, but they also have environmental effects that cannot be ignored. It is known that dielectric or electrolytic liquids used in methods such as EDM and ECM pollutes soil and water resources as waste. In addition, aerosols and gases released during processing have harmful effects on operator health and the environment. On the other hand, in these methods, which mostly depend on electrical energy, energy consumption is very high due to the machining time due to the chip removal mechanism. In this context, magnetic, ultrasonic, abrasive-assisted, or hybrid EDM technologies should be applied in order to reduce environmental damage and machining costs caused by energy consumption. In addition, the environmental sensitivity of vegetable and pure water-based dielectric fluid, dry or near-dry EDM applications has been proven by many studies. It is also recommended to add additives to prevent corrosion and improve thermal conductivity for vegetable-based dielectric fluids.

Studies have shown that each machining method has environmentally beneficial and harmful aspects according to its usage areas. At this point, more investment should be made in energy production from environmentally friendly sources, primarily to reduce harmful emissions and energy consumption, the manufacturing processes and machine tool industry should fully implement the requirements of the relevant standards to reduce energy consumption and achieve carbon efficiency. In addition, it is suggested that important cutting conditions such as tool wear rate, thermomechanical properties of the workpiece and cutting tool materials should be included in the prediction models in modeling the energy consumption in conventional and unconventional machining.

REFERENCES

Agrawal, C., Khanna, N., Gupta, M. K., & Kaynak, Y. (2020). Sustainability assessment of in-house developed environment-friendly hybrid techniques for turning Ti-6Al-4V. *Sustainable Materials and Technologies*, *26*, e00220. https://doi.org/10.1016/j.susmat.2020.e00220

Akhil, C. S., Ananthavishnu, M. H., Akhil, C. K., Afeez, P. M., Akhilesh, R., & Rahul, R. (2016). Measurement of cutting temperature during machining. *IOSR Journal of Mechanical and Civil Engineering*, *13*(2), 108–122.

Arif, U., Ali Khan, I., & Hasan, F. (2022). Green and sustainable electric discharge machining: A review. *Advances in Materials and Processing Technologies*, *9*(3), 1–75. https://doi.org/10.1080/2374068X.2022.2108599

Astakhov, V. P., & Shvets, S. (2004). The assessment of plastic deformation in metal cutting. *Journal of Materials Processing Technology*, *146*(2), 193–202. https://doi.org/10.1016/j.jmatprotec.2003.10.015

Bal, H. Ç., & Erkan, Ç. (2019). Industry 4.0 and competitiveness. *Procedia Computer Science*, *158*, 625–631. https://doi.org/10.1016/j.procs.2019.09.096

Baron, P., Dobránsky, J., Pollák, M., Cmorej, T., & Kočiško, M. (2016). Proposal of the knowledge application environment of calculating operational parameters for conventional machining technology. *Key Engineering Materials*, *669*, 95–102. https://doi.org/10.4028/www.scientific.net/KEM.669.95

Benedicto, E., Carou, D., & Rubio, E. M. (2017). Technical, economic and environmental review of the lubrication/cooling systems used in machining processes. *Procedia Engineering*, *184*, 99–116. https://doi.org/10.1016/j.proeng.2017.04.075

Bhattacharyya, B., & Doloi, B. (2019). *Modern Machining Technology: Advanced, Hybrid, Micro Machining and Super Finishing Technology*. Academic Press.

Bhowmik, S., Jagadish, & Ray, A. (2017). Abrasive water jet machining of composite materials. In K. Gupta (Ed.), *Advanced Manufacturing Technologies: Modern Machining, Advanced Joining, Sustainable Manufacturing* (pp. 77–97). Springer International Publishing. https://doi.org/10.1007/978-3-319-56099-1_4

Byrne, G., Damm, O., Monostori, L., Teti, R., van Houten, F., Wegener, K., Wertheim, R., & Sammler, F. (2021). Towards high performance living manufacturing systems—A new convergence between biology and engineering. *CIRP Journal of Manufacturing Science and Technology*, *34*, 6–21. https://doi.org/10.1016/j.cirpj.2020.10.009

Çakıroğlu, R. (2021). Machinability analysis of Inconel 718 superalloy with AlTiN-coated carbide tool under different cutting environments. *Arabian Journal for Science and Engineering, 46*(8), 8055–8073. https://doi.org/10.1007/s13369-021-05626-3

Çakıroğlu, R. (2022). Analysis of Ranque–Hilsch vortex tube cooling performance in respect of cutting temperature, resultant cutting force and chip morphology in turning of BeCu alloy. *Journal of the Brazilian Society of Mechanical Sciences and Engineering, 44*(8), 371. https://doi.org/10.1007/s40430-022-03689-3

Cetin, M. H., & Kabave Kilincarslan, S. (2020). Effects of cutting fluids with nano-silver and borax additives on milling performance of aluminium alloys. *Journal of Manufacturing Processes, 50*, 170–182. https://doi.org/10.1016/j.jmapro.2019.12.042

Chandan, P. B., Manoj, A., Gajrani, K. K., Dhaka, S., & Sankar, M. R. (2022). Sustainability issues in advanced machining processes. In P. B. Chandan, A. Manoj, K. K. Gajrani, S. Dhaka, & M. R. Sankar (Eds.), *Advanced Machining Science* (pp. 425–468). CRC Press.

Clarke, C. J., Tu, W.-C., Levers, O., Bröhl, A., & Hallett, J. P. (2018). Green and sustainable solvents in chemical processes. *Chemical Reviews, 118*(2), 747–800. https://doi.org/10.1021/acs.chemrev.7b00571

Cui, X., Guo, Y., Guo, J., & Ming, P. (2020). Bio-inspired design of cleaner interrupted turning and its effects on specific cutting energy and harmful gas emission. *Journal of Cleaner Production, 271*, 122354. https://doi.org/10.1016/j.jclepro.2020.122354

Daniyan, I., Mpofu, K., Ramatsetse, B., & Gupta, M. (2021). Review of life cycle models for enhancing machine tools sustainability: Lessons, trends and future directions. *Heliyon, 7*(4), e06790. https://doi.org/10.1016/j.heliyon.2021.e06790

Debnath, S., Reddy, M. M., & Pramanik, A. (2019). Dry and near-dry machining techniques for green manufacturing. In K. Gupta (Ed.), *Innovations in Manufacturing for Sustainability* (pp. 1–27). Springer International Publishing. https://doi.org/10.1007/978-3-030-03276-0_1

Dhanalakshmi, S., & Rameshbabu, T. (2021). Comparative study of parametric influence on wet and dry machining of LM 25 aluminium alloy. *Materials Today: Proceedings, 39*, 48–53. https://doi.org/10.1016/j.matpr.2020.06.101

Dixit, U. S., Sarma, D. K., & Davim, J. P. (2012). *Environmentally Friendly Machining.* Springer Science & Business Media.

Dornfeld, D. A. (2014). Moving towards green and sustainable manufacturing. *International Journal of Precision Engineering and Manufacturing-Green Technology, 1*(1), 63–66. https://doi.org/10.1007/s40684-014-0010-7

Falkinham, J. O. (2003). Mycobacterial aerosols and respiratory disease. *Emerging Infectious Diseases, 9*(7), 763–767. doi:10.3201/eid0907.02-0415

Gajrani, K. K., Suvin, P. S., Kailas, S. V., & Mamilla, R. S. (2019). Thermal, rheological, wettability and hard machining performance of MoS_2 and CaF_2 based minimum quantity hybrid nano-green cutting fluids. *Journal of Materials Processing Technology, 266*, 125–139. https://doi.org/10.1016/j.jmatprotec.2018.10.036

Goindi, G. S., & Sarkar, P. (2017). Dry machining: A step towards sustainable machining – Challenges and future directions. *Journal of Cleaner Production, 165*, 1557–1571. https://doi.org/10.1016/j.jclepro.2017.07.235

Groover, M. P. (2010). *Fundamentals of Modern Manufacturing* (4th ed.). John Wiley & Sons. http://10.175.10.18:8080/jspui/handle/123456789/690

Gunjal, S. U., & Patil, N. G. (2018). Experimental investigations into turning of hardened AISI 4340 Steel using vegetable based cutting fluids under minimum quantity lubrication. *Procedia Manufacturing, 20*, 18–23. https://doi.org/10.1016/j.promfg.2018.02.003

Günan, F., Kıvak, T., Yıldırım, Ç. V., & Sarıkaya, M. (2020). Performance evaluation of MQL with Al_2O_3 mixed nanofluids prepared at different concentrations in milling of Hastelloy C276 alloy. *Journal of Materials Research and Technology, 9*(5), 10386–10400. https://doi.org/10.1016/j.jmrt.2020.07.018.

Günay, M., Korkmaz, M. E., & Yaşar, N. (2020). Performance analysis of coated carbide tool in turning of Nimonic 80A superalloy under different cutting environments. *Journal of Manufacturing Processes*, *56*, 678–687. https://doi.org/10.1016/j.jmapro.2020.05.031

Gupta, K. (2020). A review on green machining techniques. *Procedia Manufacturing*, *51*, 1730–1736. https://doi.org/10.1016/j.promfg.2020.10.241

Gupta, M. K., Jamil, M., Wang, X., Song, Q., Liu, Z., Mia, M., Hegab, H., Khan, A. M., Collado, A. G., Pruncu, C. I., & Imran, G. M. S. (2019). Performance evaluation of vegetable oil-based nano-cutting fluids in environmentally friendly machining of Inconel-800 alloy. *Materials*, *12*(17), Article 17. https://doi.org/10.3390/ma12172792

Hegab, H., Kishawy, H. A., Gadallah, M. H., Umer, U., & Deiab, I. (2018a). On machining of Ti-6Al-4V using multi-walled carbon nanotubes-based nano-fluid under minimum quantity lubrication. *The International Journal of Advanced Manufacturing Technology*, *97*(5), 1593–1603. https://doi.org/10.1007/s00170-018-2028-4

Hegab, H., Umer, U., Deiab, I., & Kishawy, H. (2018b). Performance evaluation of Ti–6Al–4V machining using nano-cutting fluids under minimum quantity lubrication. *The International Journal of Advanced Manufacturing Technology*, *95*(9), 4229–4241. https://doi.org/10.1007/s00170-017-1527-z

Hoghoughi, M. H., Farahnakian, M., & Elhami, S. (2022). Environmental, economical, and machinability based sustainability assessment in hybrid machining process employing tool textures and solid lubricant. *Sustainable Materials and Technologies*, *34*, e00511. https://doi.org/10.1016/j.susmat.2022.e00511

Jain, A., & Kansal, H. (2017). Green machining–machining of the future. *4th National Conference on Advancements in Simulation & Experimental Techniques in Mechanical Engineering (NCASEme-2017)*, March, 21–25.

Jamil, M., Zhao, W., He, N., Gupta, M. K., Sarikaya, M., Khan, A. M., Sanjay, M. R., Siengchin, S., & Pimenov, D. Y. (2021). Sustainable milling of Ti–6Al–4V: A trade-off between energy efficiency, carbon emissions and machining characteristics under MQL and cryogenic environment. *Journal of Cleaner Production*, *281*, 125374. https://doi.org/10.1016/j.jclepro.2020.125374

Jerman, M., Zeleňák, M., Lebar, A., Foldyna, V., Foldyna, J., & Valentinčič, J. (2021). Observation of cryogenically cooled ice particles inside the high-speed water jet. *Journal of Materials Processing Technology*, *289*, 116947. https://doi.org/10.1016/j.jmatprotec.2020.116947

Jia, S., Yuan, Q., Lv, J., Liu, Y., Ren, D., & Zhang, Z. (2017). Therblig-embedded value stream mapping method for lean energy machining. *Energy*, *138*, 1081–1098. https://doi.org/10.1016/j.energy.2017.07.120

Kalyon, A., & Fatatit, A. Y. (2019). The environmental impact of electric discharge machining. *International Journal of Engineering Science and Application*, *3*(3), Article 3.

Kazeem, R. A., Fadare, D. A., Ikumapayi, O. M., Adediran, A. A., Aliyu, S. J., Akinlabi, S. A., Jen, T.-C., & Akinlabi, E. T. (2022). Advances in the application of vegetable-oil-based cutting fluids to sustainable machining operations—A review. *Lubricants*, *10*(4), Article 4. https://doi.org/10.3390/lubricants10040069

Khan, A. M., Jamil, M., Mia, M., Pimenov, D. Y., Gasiyarov, V. R., Gupta, M. K., & He, N. (2018). Multi-objective optimization for grinding of AISI D2 steel with Al_2O_3 wheel under MQL. *Materials*, *11*(11), Article 11. https://doi.org/10.3390/ma11112269

Khan, A. M., Jamil, M., Mia, M., He, N., Zhao, W., & Gong, L. (2020). Sustainability-based performance evaluation of hybrid nanofluid assisted machining. *Journal of Cleaner Production*, *257*, 120541. https://doi.org/10.1016/j.jclepro.2020.120541

Khanna, N., Shah, P., Wadhwa, J., Pitroda, A., Schoop, J., & Pusavec, F. (2021). Energy consumption and lifecycle assessment comparison of cutting fluids for drilling titanium alloy. *Procedia CIRP*, *98*, 175–180. https://doi.org/10.1016/j.procir.2021.01.026

Kiran, Tomar, H., & Gupta, N. (2021). Sustainability concerns of non-conventional machining processes—An exhaustive review. In R. Agrawal, J. K. Jain, V. S. Yadav, V. K. Manupati, & L. Varela (Eds.), *Recent Advances in Smart Manufacturing and Materials* (pp. 263–273). Springer. https://doi.org/10.1007/978-981-16-3033-0_25

Kong, T., Nhg, D. R., & Shi, V. T. (2022). Method for optimizing energy consumption in machining manufacturing process. *Mathematical Problems in Engineering, 2022,* e8300666. https://doi.org/10.1155/2022/8300666

Korkmaz, M. E., Gupta, M. K., Krolczyk, G. M., Maruda, R. W., & Li, Z. (2021). Effect of nanoparticles as a lubricants in nano-MQL machining of metallic materials: A review. *2021 6th International Conference on Nanotechnology for Instrumentation and Measurement (NanofIM),* 1–7. https://doi.org/10.1109/NanofIM54124.2021.9737354

Kui, G. W. A., Islam, S., Reddy, M. M., Khandoker, N., & Chen, V. L. C. (2022). Recent progress and evolution of coolant usages in conventional machining methods: A comprehensive review. *The International Journal of Advanced Manufacturing Technology, 119*(1), 3–40. https://doi.org/10.1007/s00170-021-08182-0

Kumar Gajrani, K., & Ravi Sankar, M. (2017). Past and current status of eco-friendly vegetable oil based metal cutting fluids. *Materials Today: Proceedings, 4*(2, Part A), 3786–3795. https://doi.org/10.1016/j.matpr.2017.02.275

Kumar Sharma, A., Kumar Tiwari, A., Rai Dixit, A., & Kumar Singh, R. (2020). Measurement of machining forces and surface roughness in turning of AISI 304 steel using alumina-MWCNT hybrid nanoparticles enriched cutting fluid. *Measurement, 150,* 107078. https://doi.org/10.1016/j.measurement.2019.107078

Kuram, E., Ozcelik, B., & Demirbas, E. (2013). Environmentally friendly machining: Vegetable based cutting fluids. In J. P. Davim (Ed.), *Green Manufacturing Processes and Systems* (pp. 23–47). Springer. https://doi.org/10.1007/978-3-642-33792-5_2

Liang, X., Liu, Z., Chen, L., Hao, G., Wang, B., Cai, Y., & Song, Q. (2020). Tool wear induced modifications of plastic flow and deformed material depth in new generated surfaces during turning Ti-6Al-4V. *Journal of Materials Research and Technology, 9*(5), 10782–10795. doi:10.1016/j.jmrt.2020.07.093.

Luo, H., Wang, Y., & Zhang, P. (2020). Effect of cutting and vibration parameters on the cutting performance of 7075-T651 aluminum alloy by ultrasonic vibration. *The International Journal of Advanced Manufacturing Technology, 107*(1), 371–384. https://doi.org/10.1007/s00170-020-05098-z

Lv, J., Tang, R., Tang, W., Jia, S., Liu, Y., & Cao, Y. (2018). An investigation into methods for predicting material removal energy consumption in turning. *Journal of Cleaner Production, 193,* 128–139. https://doi.org/10.1016/j.jclepro.2018.05.035

Maier, H. J., Herbst, S., Denkena, B., Dittrich, M.-A., Schaper, F., Worpenberg, S., Gustus, R., & Maus-Friedrichs, W. (2020). Towards dry machining of titanium-based alloys: A new approach using an oxygen-free environment. *Metals, 10*(9), Article 9. https://doi.org/10.3390/met10091161

Melentiev, R., & Fang, F. (2018). Recent advances and challenges of abrasive jet machining. *CIRP Journal of Manufacturing Science and Technology, 22,* 1–20. https://doi.org/10.1016/j.cirpj.2018.06.001

Perec, A. (2021). Research into the disintegration of abrasive materials in the abrasive water jet machining process. *Materials, 14*(14), Article 14. https://doi.org/10.3390/ma14143940

Pervaiz, S., Kannan, S., Deiab, I., & Kishawy, H. (2020). Role of energy consumption, cutting tool and workpiece materials towards environmentally conscious machining: A comprehensive review. *Proceedings of the Institution of Mechanical Engineers, Part B: Journal of Engineering Manufacture, 234*(3), 335–354. https://doi.org/10.1177/0954405419875344

Pimenov, D. Y., Mia, M., Gupta, M. K., Machado, Á. R., Pintaude, G., Unune, D. R., Khanna, N., Khan, A. M., Tomaz, Í., Wojciechowski, S., & Kuntoğlu, M. (2022). Resource saving by optimization and machining environments for sustainable manufacturing: A review and future prospects. *Renewable and Sustainable Energy Reviews, 166*, 112660. https://doi.org/10.1016/j.rser.2022.112660

Radosz, J. (2020). Ultrasonic noise measurements in the work environment. In D. Pleban (Ed.), *Occupational Noise and Workplace Acoustics* (221–251). CRC Press.

Rajesh, S., Gobikrishnan, U., Krishnarjuna Rao, N., Balamurugan, R., Senthilkumar, K. M., Selvan, T. A., & Madhankumar, S. (2022). Electrochemical machining of aluminium 7075 alloy, silicon carbide, and fly ash composites: An experimental investigation of the effects of variables on material removal rate. *Materials Today: Proceedings, 62*, 863–867. https://doi.org/10.1016/j.matpr.2022.04.054

Ramachandran, K., Yeesvaran, B., Kadirgama, K., Ramasamy, D., Ghani, S. A. C., & Anamalai, K. (2017). State of art of cooling method for dry machining. *MATEC Web of Conferences, 90*, 01015. https://doi.org/10.1051/matecconf/20179001015

Rohith, R., Shreyas, B. K., Kartikgeyan, S., Sachin, B. A., Umesha, K., & Nanjundeswaraswamy, D. T. S. (2019). Selection of non-traditional machining process. *International Journal of Engineering Research & Technology, 8*(11), 148–155.

Ross, N. S., Mia, M., Anwar, S. G. M., Saleh, M., & Ahmad, S. (2021). A hybrid approach of cooling lubrication for sustainable and optimized machining of Ni-based industrial alloy. *Journal of Cleaner Production, 321*, 128987. https://doi.org/10.1016/j.jclepro.2021.128987

Salguero, J., Vazquez-Martinez, J. M., Sol, I. D., & Batista, M. (2018). Application of pin-on-disc techniques for the study of tribological interferences in the dry machining of A92024-T3 (Al–Cu) alloys. *Materials, 11*(7), Article 7. https://doi.org/10.3390/ma11071236

Sankaranarayanan, R., Rajesh Jesudoss Hynes, N., Senthil Kumar, J., & Krolczyk, G. M. (2021). A comprehensive review on research developments of vegetable-oil based cutting fluids for sustainable machining challenges. *Journal of Manufacturing Processes, 67*, 286–313. https://doi.org/10.1016/j.jmapro.2021.05.002

Saptaji, K., Gebremariam, M. A., & Azhari, M. A. B. M. (2018). Machining of biocompatible materials: A review. *The International Journal of Advanced Manufacturing Technology, 97*(5), 2255–2292. https://doi.org/10.1007/s00170-018-1973-2

Sen, B., Mia, M., Krolczyk, G. M., Mandal, U. K., & Mondal, S. P. (2021). Eco-friendly cutting fluids in minimum quantity lubrication assisted machining: A review on the perception of sustainable manufacturing. *International Journal of Precision Engineering and Manufacturing-Green Technology, 8*(1), 249–280. https://doi.org/10.1007/s40684-019-00158-6

Sharma, A. K., Tiwari, A. K., & Dixit, A. R. (2016). Effects of minimum quantity lubrication (MQL) in machining processes using conventional and nanofluid based cutting fluids: A comprehensive review. *Journal of Cleaner Production, 127*, 1–18. https://doi.org/10.1016/j.jclepro.2016.03.146

Shokrani, A., Dhokia, V., & Newman, S. T. (2016). Investigation of the effects of cryogenic machining on surface integrity in CNC end milling of Ti–6Al–4V titanium alloy. *Journal of Manufacturing Processes, 21*, 172–179. https://doi.org/10.1016/j.jmapro.2015.12.002

Siju, A. S., Gajrani, K. K., & Joshi, S. S. (2021). Dual textured carbide tools for dry machining of titanium alloys. *International Journal of Refractory Metals and Hard Materials, 94*, 105403. https://doi.org/10.1016/j.ijrmhm.2020.105403

Singh, B., Sasi, R., Kanmani Subbu, S., & Muralidharan, B. (2019). Electric discharge texturing of HSS cutting tool and its performance in dry machining of aerospace alloy. *Journal of the Brazilian Society of Mechanical Sciences and Engineering, 41*(3), 152. https://doi.org/10.1007/s40430-019-1654-6

Singh, J., & Sharma, R. (2017). Green EDM strategies to minimize environmental impact and improve process efficiency. *Journal for Manufacturing Science and Production, 16.* https://doi.org/10.1515/jmsp-2016-0034

Singh, R., & Sharma, V. (2022). Recent trends of cutting fluids and lubrication techniques in machining. In Y. Singh, N. K. Singh, & M. Ram (Eds.), *Advanced Manufacturing Processes* (1–28). CRC Press.

Singh, T., Dureja, J. S., Dogra, M., & Bhatti, M. S. (2018). Environment friendly machining of Inconel 625 under nano-fluid minimum quantity lubrication (NMQL). *International Journal of Precision Engineering and Manufacturing, 19*(11), 1689–1697. https://doi.org/10.1007/s12541-018-0196-7

Singh, T., Singh, P., Dureja, J. s., Dogra, M., Singh, H., & Bhatti, M. S. (2016). A review of near dry machining/minimum quantity lubrication machining of difficult to machine alloys. *International Journal of Machining and Machinability of Materials, 18*(3), 213–251. https://doi.org/10.1504/IJMMM.2016.076276

Song, K. H., Lim, D. W., Park, J. Y., Ha, S. J., & Yoon, G. S. (2020). Investigation on influence of hybrid nozzle of CryoMQL on tool wear, cutting force, and cutting temperature in milling of titanium alloys. *The International Journal of Advanced Manufacturing Technology, 110*(7), 2093–2103. https://doi.org/10.1007/s00170-020-05646-7

Sreejith, P. S., & Ngoi, B. K. A. (2000). Dry machining: Machining of the future. *Journal of Materials Processing Technology, 101*(1), 287–291. https://doi.org/10.1016/S0924-0136(00)00445-3

Stanojković, J.; Radovanović, M. (2022). Influence of the cutting parameters on force, moment and surface roughness in the end milling of aluminum 6082-T6. *Facta Universitatis, Series: Mechanical Engineering (FU Mech Eng), 20*(1), 157–165. https://doi.org/10.22190/FUME180220002S

Tang, L., Zhang, Y., Li, C., Zhou, Z., Nie, X., Chen, Y., Cao, H., Liu, B., Zhang, N., Said, Z., Debnath, S., Jamil, M., Ali, H. M., & Sharma, S. (2022). Biological stability of water-based cutting fluids: Progress and application. *Chinese Journal of Mechanical Engineering, 35*(1), 3. https://doi.org/10.1186/s10033-021-00667-z

Teti, R., D'Addona, D. M., & Segreto, T. (2021). Microbial-based cutting fluids as bio-integration manufacturing solution for green and sustainable machining. *CIRP Journal of Manufacturing Science and Technology, 32*, 16–25. https://doi.org/10.1016/j.cirpj.2020.09.016

Uddin, G. M., Joyia, F. M., Ghufran, M., Khan, S. A., Raza, M. A., Faisal, M., Arafat, S. M., Zubair, S. W. H., Jawad, M., Zafar, M. Q., Irfan, M., Waseem, B., Chaudhry, I. A., & Zeid, I. (2021). Comparative performance analysis of cemented carbide, TiN, TiAlN, and PCD coated inserts in dry machining of Al 2024 alloy. *The International Journal of Advanced Manufacturing Technology, 112*(5), 1461–1481. https://doi.org/10.1007/s00170-020-06315-5

Wang, X., Li, C., Zhang, Y., Ding, W., Yang, M., Gao, T., Cao, H., Xu, X., Wang, D., Said, Z., Debnath, S., Jamil, M., & Ali, H. M. (2020). Vegetable oil-based nanofluid minimum quantity lubrication turning: Academic review and perspectives. *Journal of Manufacturing Processes, 59*, 76–97. https://doi.org/10.1016/j.jmapro.2020.09.044

Wang, Y., Li, C., Zhang, Y., Yang, M., Li, B., Dong, L., & Wang, J. (2018). Processing characteristics of vegetable oil-based nanofluid MQL for grinding different workpiece materials. *International Journal of Precision Engineering and Manufacturing-Green Technology, 5*(2), 327–339. https://doi.org/10.1007/s40684-018-0035-4

Williams, J. G., Patel, Y., & Blackman, B. R. K. (2010). A fracture mechanics analysis of cutting and machining. *Engineering Fracture Mechanics, 77*(2), 293–308. https://doi.org/10.1016/j.engfracmech.2009.06.011

Yap, T. C. (2019). Roles of cryogenic cooling in turning of superalloys, ferrous metals, and viscoelastic polymers. *Technologies, 7*(3), Article 3. https://doi.org/10.3390/technologies7030063

Yi, S., Li, N., Solanki, S., Mo, J., & Ding, S. (2019). Effects of graphene oxide nanofluids on cutting temperature and force in machining Ti-6Al-4V. *The International Journal of Advanced Manufacturing Technology, 103*(1), 1481–1495. https://doi.org/10.1007/s00170-019-03625-1

Yusuf, L. A., Popoola, K., & Musa, H. (2021). A review of energy consumption and minimisation strategies of machine tools in manufacturing process. *International Journal of Sustainable Engineering, 14*(6), 1826–1842. https://doi.org/10.1080/19397038.2021.1964633

Zahoor, S., Ameen, F., Abdul-Kader, W., & Stagner, J. (2020). Environmentally conscious machining of Inconel 718: Surface roughness, tool wear, and material removal rate assessment. *The International Journal of Advanced Manufacturing Technology, 106*(1), 303–313. https://doi.org/10.1007/s00170-019-04550-z

Zhang, Z., Cui, J., Zhang, J., Liu, D., Yu, Z., & Guo, D. (2019). Environment friendly chemical mechanical polishing of copper. *Applied Surface Science, 467–468*, 5–11. https://doi.org/10.1016/j.apsusc.2018.10.133

Zhao, F., & Sharma, A. (2015). *Environmentally Friendly Machining Machining Process Environmentally Friendly Machining* (pp. 1127–1154). https://doi.org/10.1007/978-1-4471-4670-4_14

Zheng, J., Zheng, W., Chen, A., Yao, J., Ren, Y., Zhou, C., Wu, J., Ling, W., Bai, B., Wang, W., & Zhang, Z. (2020). Sustainability of unconventional machining industry considering impact factors and reduction methods of energy consumption: A review and analysis. *Science of The Total Environment, 722*, 137897. https://doi.org/10.1016/j.scitotenv.2020.137897

Zhong, Q., Tang, R., & Peng, T. (2017). Decision rules for energy consumption minimization during material removal process in turning. *Journal of Cleaner Production, 140*, 1819–1827. https://doi.org/10.1016/j.jclepro.2016.07.084

Index

abrasive powder 61
abrasive water jet machining 17, 32, 38, 53–54, 79, 88, 92, 170, 174, 177
airborne 10–11, 14
Al2O3 30, 76
allergic 11, 107
AlSiC 66
amplitude 18, 22, 85–86
antioxidant 65, 71, 166
aquatic ecosystem 9
AZ31B 68, 156
B_4C 95, 96, 113
bags 12
benchmarking 7
biocide 8, 107, 144
biodegradable 2, 11–12, 15, 62–63, 65, 73, 107, 117, 124, 144, 149, 166
biodiesel 63, 68, 70–71, 74, 77
boring 7
boron 9, 16, 80
building 3, 15, 102
built-up edge 7, 104, 117, 148, 165
canola oil 13, 65, 68, 167
carbide 8, 33, 78, 81–83, 89, 93–94, 110, 126, 145, 165, 169, 173, 175–176, 178–179
carbon-carbon composite 145
carbon nanotube 110, 126, 141, 176
carcinogenic 11, 107
cartridge 12
case 16, 35, 37, 51, 57, 65, 78–79, 90, 97, 111–112, 123, 126, 128–129, 135–136, 138, 141, 160, 168, 170, 176
cement 3
ceramic 13, 15, 21, 25, 28, 31–32, 75, 77, 79, 80–84, 86–87, 89–94, 145, 149
ceramic matrix composite 75, 81, 91, 93, 145
CH_4 62
chip 7, 16, 26, 69, 78–79, 81, 83, 87, 97, 103–107, 112, 116–117, 132, 135–136, 139–140, 148, 150, 155, 158, 164–165, 168, 171, 173, 175
circular economy 4, 118, 154
coalescer 13
cobalt 9, 22, 37
coconut 15, 64–65, 68–69, 71, 80, 94, 153, 162, 167
complex connectivity 4
composite material 50, 72, 74–75, 145, 147–148, 154–156, 174
conventional machining 28, 35, 38–39, 54, 79, 84, 86, 88, 90, 116, 144, 158, 165, 171–174

coolant 2, 7–13, 16, 49, 79, 87, 104, 106–112, 117–119, 122–123, 137, 144, 148–149, 151, 158, 167, 169, 171–172, 177
coolant-based 12
corrosion 7, 30, 33, 35, 37–39, 53, 55, 63, 69, 105, 111, 129, 135, 144–145, 149–150, 153, 162, 167, 170–171, 174
cryogenic cooling 74, 79, 83, 90–91, 110, 113, 121, 123, 131–132, 135–138, 140–142, 151, 156, 162, 164–165, 168, 179
CryoMQL 114, 179
cubical block 7
cutting fluid 1, 4, 7–8, 10–16, 79–80, 90, 93, 105, 107–110, 112, 114–118, 120, 122–123, 126–127, 137, 144–145, 148, 153–154, 156, 158–156, 165–169, 171–173, 175–179
cutting force 19, 21–22, 28, 54, 88, 105, 110–111, 114, 116–117, 121, 124, 127, 129, 131, 141, 147–148, 150, 152, 154–155, 161–162, 165–166, 168–169, 172, 175, 179
cutting insert 8
cutting speed 20, 44, 46, 78, 102–106, 108, 112, 115, 148, 162, 167–168
cylindrical bar 7
deformation zone 106–107
delamination 46–47, 49, 91, 147, 150–152, 155–156
depth of cut 78, 102, 104, 148, 167
diamond 8, 34, 37, 115, 148
diamond-like carbon 37
dielectric fluid 40–41, 53–54, 60, 62, 66, 70–77, 81, 84, 144, 153, 156, 170, 174
die-sinking 56–57, 66–69
difficult-to-machine 19–20, 22, 25, 27, 35, 48, 78, 120–121, 154
discharge current 55, 58, 62, 68–69
drilling 7, 15–19, 21–23, 25–21, 33–34, 40, 44, 46, 48–49, 51–52, 66, 72, 75, 86–87, 90, 92, 95, 117, 123–124, 142, 146–147, 150–152, 156, 158, 173, 176
droplet 10–11, 79, 82, 85, 114, 167
dry cutting 83, 91–92, 108, 110, 112, 121–123, 164, 166
dry machining 11, 80, 90–91, 94, 108, 111, 115–117, 119–114, 138, 141–142, 145, 148–149, 152–155, 162–163, 168–170, 173, 175, 177–179
dust 8, 10–11, 68, 86, 140, 155

eco-friendly 5, 7, 16, 54, 61–62, 70–71, 74, 76–77, 93, 116, 119, 121, 142, 153, 177–178
electric discharge machining 17, 39, 48–49, 53, 72–73, 75–77, 92, 144, 154, 174, 176
electrochemical machining 2, 30–31, 34, 39, 54, 171, 174
electrode wire 39, 41
electronic 3, 44
electrostatic 54, 112
emission 2, 5, 7, 10–11, 14, 62, 65–71, 73, 97, 99, 101, 113, 144, 152, 154, 157, 165, 169, 170, 172, 174–176
energy 2, 4, 6, 8–10, 13–15, 17, 20, 22, 32, 38–41, 44, 46, 48, 54–56, 58, 60–63, 65–70, 72, 74–75, 84, 90, 92–93, 96, 98–100, 101–102, 108, 111–115, 118–119, 122, 129, 140, 143, 157–158, 160, 162, 164, 168–180
energy and water usage 4
energy consumption 2, 10, 62, 66, 69–70, 75, 96, 98, 101–102, 112, 115, 118, 143, 157, 160, 162, 170, 172–174, 176–177, 180
environmental damage 3, 90, 157, 165, 173–174
environmental effect 7–8, 12, 14, 63, 66, 68, 97, 99–101, 110–111, 153, 158–150, 162, 164, 171, 173
environmentally benign machining 1–2, 11, 143, 148, 153, 157, 170, 172
feed rate 11, 39, 42, 45, 48, 78, 89, 102, 104, 112, 148, 168
ferrous 3, 179
filtration 12–13, 159
finishing process 17, 54
flash point 63–67, 70, 167
flow rate 13, 66, 68–69, 82, 89, 151
forklift 10
fossil fuel 3, 172
frequency 22, 45–46, 85, 87
friction 7, 9, 39, 78, 80, 81, 91, 106, 108, 114, 123, 144, 158, 162–164, 166–167, 169, 173
graphite tool 68
green machining 35, 78–79, 90–91, 117, 121–123, 137–138, 141, 157, 176
greenhouse gas 2–3, 62, 101, 143
grinding 7, 11, 14, 17–19, 21–23, 25–32, 56, 81–82, 90, 92–93, 120, 123–124, 146, 150, 158, 166, 169, 171, 176, 179
harmful gas 62–63, 65, 171–172, 175
harmful substance 9–10, 69, 79
hazardous material 1, 10, 143
heat 7, 9–10, 17–22, 24–25, 29–30, 32–33, 36–39, 43–46, 48, 52, 56, 58–61, 63–67, 78, 80–81, 83–84, 87, 90, 93, 105–109, 114–115, 117–118, 122–123, 126–127, 132, 141, 144, 147, 151, 153, 155, 158, 160–161, 163, 167–169

heat affected zone 25, 56, 58
heat assisted machining 17, 19, 29–30, 32, 83
heat capacity 36, 65
heat generation 19, 78, 81, 90, 105–106, 126, 147, 153
heat resistance 155
helium 63, 68–69, 77, 122, 131, 164
high-pressure 41, 44, 79, 105, 108, 123, 140, 146
high speed 33, 48, 84, 90, 140, 167, 170
hybrid composite 71, 146, 156
hybrid machining 17–19, 28–29, 31, 149, 155, 176
hybrid waterjet machining 17, 19–20, 27, 29
hydraulic oil 8
hydrocarbon 57, 60, 62, 70, 81, 107, 158
Inconel 16, 30, 32–34, 46, 68–69, 72–73, 94, 118, 120, 123, 128, 141–142, 167, 169, 175–176, 179–180
incineration 3
industrial ecology framework 3
insulation 15, 27, 59–61, 63, 65, 74–75
jatropha oil 13, 16, 66, 68, 76
jet 17, 19, 27–33, 38, 44, 52, 54, 79, 88–89, 91–92, 94, 113, 122, 141, 146, 160, 170–171, 174, 176
kerf geometry 89, 94
kerf taper 88–89
kerosene 40, 62, 64, 66–68
landfilling 3, 11
laser 17–21, 24–25, 28, 30–35, 38–39, 41, 44–54, 79, 81, 83–84, 87–94, 146, 154, 160, 170
laser assisted machining 33
laser beam machining 21, 35, 39, 44, 48–49, 51, 79, 87, 90–91, 93
laser current 46
laser machining 17, 19, 38, 45, 46, 48–50, 54, 94
laser power 44, 45, 47, 81, 84, 88
laser wavelength 44, 45
life cycle 3–6, 14–16, 96–102, 111–114, 116–120, 156, 175
liquid nitrogen 14, 82, 113, 131–132, 151, 155
liquid petroleum gas 24
low thermal conductivity 39, 46, 78
lubricant 2, 7–8, 10, 14, 16, 79–81, 83, 91, 105, 109, 123, 132, 142, 149, 153–154, 156, 158, 164, 167, 169, 176–177
machine tool 8, 14, 30, 33–34, 49–51, 72–76, 91–92, 94–95, 103, 116, 120, 142, 172–175, 180
machining process 1–2, 7–11, 14, 16–21, 28–32, 35, 38–40, 42, 44, 47–48, 50–51, 54–55, 59, 71–73, 75, 77, 79, 92–93, 96, 102, 105, 108–110, 112–113, 115–118, 120, 122, 124, 141, 156, 159, 172, 174, 175–178, 180
machining quality 57, 59, 69, 86, 88, 169

manufacturing 1–8, 11, 13–17, 19–20, 22, 27, 30–34, 38, 48–54, 71–79, 86, 89–94, 96–97, 99, 101–102, 105, 116–123, 128, 139, 140–144, 146, 148, 152, 154, 157–160, 163, 164, 168, 170–172, 174–180
mass production 2, 19, 62
material cycle 3–4, 116
material removal rate 19, 21, 22, 49, 51, 69, 72, 74, 114, 134, 170, 178, 180
material thickness 44
material transformation 5
matrix method 4, 50
membrane microfiltration 13
metal 4, 7, 11, 15–16, 19, 30, 33, 38, 50–51, 58–59, 71–72, 78, 81, 92, 94, 105, 109–111, 116, 121, 123, 126–127, 141, 144–146, 149, 150, 155, 164, 170–171, 174, 177
metal matrix composite 15, 30, 33, 71, 72, 92, 94, 145, 149, 150, 155
microhardness 37, 42–44, 67–68, 135, 166
microorganism 10, 13, 111
milling 7, 11, 17–33, 40, 83–84, 87, 92, 94, 111, 116, 118, 123–124, 129, 141–142, 146, 150, 154, 156, 158, 164, 166–169, 175–176, 178–179
minerals 3, 62
mining pesticide 3
mist 8, 10–11, 30, 52, 71, 77, 82, 124, 154
molten electrode 55–56
MQL 14, 16, 79–82, 90–93, 105, 108–129, 131, 135–142, 145, 150, 152, 162–170, 173, 175–179
mustard oil 68
NanoMQL 105, 108, 110, 121–123, 128–129, 131, 136, 138–140
nanoparticle 80, 121, 130, 166, 168–169
natural resource 2, 9, 11, 97, 111, 143, 154, 164, 172–173
natural water resource 3
near-dry 16, 63, 66, 70–74, 77, 79, 120, 170, 174–175
net-shape 4
nickel 4, 9, 16, 34–35, 37, 46, 50–51, 62, 94, 126
NiTi alloy 35–40, 42, 43, 47–48
non-ferrous metal 51, 141
non-toxic 15
oil(s) 2, 8–9, 13–16, 29, 62–63, 65–68, 70–71, 75–76, 79, 83, 103, 107, 120–121, 127, 144, 152–153, 162–164, 166–168, 173
oil-based cutting fluid 13, 16, 19, 118, 154, 162, 163, 176
oil sorbent 13
open circuit voltage 58

oxidation 25, 63, 65–67, 71, 127, 153, 166, 168
oxygen 10, 24, 42, 56, 63, 66, 69–70, 74, 164, 168, 177
packaging 4, 7, 16, 101
palm 13, 64, 66, 68, 76, 153–154, 166, 169
palm oil 68, 76, 153, 166, 169
peak current 42, 46, 48
pellet stock 7
perpetual 13
pesticide 3, 5
philosophical 5
planning 12, 48, 100, 117–118, 158, 173
polarity 58, 81
pollutant 1, 9–10, 12, 14, 65, 97, 100
polymer matrix composite 145, 150, 154
polymeric 13
product flexibility 6
product quality 61, 78, 121, 134, 140, 169, 173
production stage 3
pulse duration 44–48, 58, 68–69, 87
pulse interval 55, 58, 68–69
recast layer 30, 42–44, 46–48, 52, 67–68, 81, 85, 90, 93
recycle 3, 12, 171
remanufacturing 3, 6, 11, 14, 32
renewable 63, 67, 74–75, 90, 100, 119, 140, 153, 172, 178
resource consumption 8–9, 13, 97, 143, 157
rice bran 63–65, 68, 71
rotary ultrasonic machining 86, 92–93, 95, 156
scanning speed 44, 46, 48, 87
segregating waste 11
service life 10, 162
sesame 65
Si3N4 93–94
SiC 30, 32, 69, 78, 81–82, 84, 87, 92–94, 126, 145, 150, 155
silicon carbide 89, 93, 94, 126, 145, 178
skimmers 13
smart city 102
societal impact 1–2
solid waste 3, 11, 15, 97
spark off time 41, 43, 48
spark on time 41–44, 48
specific heat 56, 61, 64, 169
steel 5, 8, 15–16, 28, 30, 32–33, 37–38, 50, 52, 62, 66–69, 72, 76–77, 91, 105, 112, 115–157, 119–120, 140–141, 154, 156, 166–168, 175–177
stock metal 4
sunflower oil 15, 65, 67–68, 75, 153
superalloy 16, 22, 24–25, 27, 29, 30, 33, 46, 49, 52, 94, 114, 123, 129, 136, 142, 164, 175, 179
surface morphology 31, 49, 58, 66, 92, 156

surface quality 8, 18–19, 21–24, 30, 32, 42, 54, 59, 61, 63, 65–66, 70, 80–81, 83–84, 88, 90, 97, 104, 108, 110–111, 113–114, 121, 123–124, 127, 132, 137, 148, 150, 152, 163–164, 166, 168–169, 171, 173
surface roughness 15, 50, 57–58, 61, 67, 71–72, 80, 82, 84–85, 87, 89, 90–92, 111, 114, 116, 126, 129–132, 135–136, 138, 141, 144, 148, 154, 158, 167–169, 172, 177, 179–180
surface texture 102, 169
surfactant 13, 33, 65, 68, 128, 142, 153
sustainability 1–2, 7, 9, 14, 16–17, 19, 35, 55, 66, 70, 75, 77, 96, 100–102, 114–115, 118–119, 134, 140, 143, 149, 154–157, 160, 164–165, 170–172, 174–177, 180
swarf 8, 11, 144
tantalum 9, 37
thermal conductivity 36, 39, 46, 61, 63–66, 68, 78, 80, 110, 121, 127, 129, 140, 141, 145, 162, 167–169, 174
thermal dissipation 65
thermo-electric 38–39, 84, 168
thread cutting 7
Ti6Al4V 31, 33, 66, 68, 71, 113, 142, 168–169, 173
TiAlN 88, 169, 179
TiB2 78
titanium alloy 15, 30, 46, 49, 52, 91, 113, 116–117, 141, 176
tool geometry 104, 106, 114, 148
tool insert 9, 151
tool material 31, 59, 62, 65, 70, 105, 108, 110, 158–159, 162, 173–174
tool wear 16, 21–22, 24–26, 33, 39, 54–55, 57–58, 60, 62–63, 68–70, 75, 78, 81–84, 87–88, 91, 95, 104–105, 110, 115–116, 118–119, 121, 124, 126, 131–132, 134–136, 138, 142, 147–150, 153–154, 156, 162, 164–165, 167–169, 171–172, 174, 177, 179–180

toughness 35, 145
toxic substance 9, 159
transverse speed 88
tungsten 9, 22, 39, 66–67, 78, 81
turning 7, 11–12, 15, 16–17, 19, 21–33, 40, 49, 81–84, 90–92, 94, 97, 103–104, 108, 112, 114, 116–117, 119–120, 123–124, 140–142, 146, 150, 154–156, 158, 165, 167–169, 173–177, 179–180
ultrasonic vibration machining 17
unconventional machining 54–55, 158, 170, 172–174, 180
urban area 3
vegetable-based cutting 2, 15–16, 90, 153, 168
vegetable dielectric 67–68, 70
vegetable oil 13, 16, 40, 53, 62–63, 65–67, 70–71, 73, 75–77, 83, 118, 120, 126, 152–153, 156, 162–163, 166–168, 173, 176–177, 179
vibration assisted machining 85, 122
viscosity 61, 63–70, 153, 168–169
volatile organic 8, 10
washing pressure 58, 63
waste 1–6, 9–11, 13, 15, 16, 53, 55, 62, 63, 65, 69, 71, 76, 86, 97–98, 100–102, 108, 112, 116, 143–144, 158–160, 165, 169–171, 173
waste management method 3
waste water 9, 101
water pollution 9, 140, 157
white layer thickness 42, 55, 68
wire material 42, 48
wire tension 39, 42, 48, 59, 70
wire wear rate 59, 69
workpiece material 12, 18, 26, 28, 38, 55, 59, 84, 104, 171, 177, 179
zero-emission 7
ZrB2 78
ZrN 78
ZrO2 76, 78, 81–82, 84, 90, 126